C000321173

THE BECOMING

ROBERT FRIPP

THE BECOMING

Notes on the
Evolution of a Small Planet

with a foreword by
JOHN FOWLES

ARTHUR JAMES
BERKHAMSTED

First published in Great Britain by

ARTHUR JAMES LTD
40 Lower Kings Road, Berkhamsted, Herts HP4 2AA

A catalogue record for this book is available
from the British Library

ISBN 0 85305 473 8

Typeset in Monotype Bulmer by
Strathmore Publishing Services, London N7

Printed in Great Britain by
Guernsey Press Ltd, Guernsey, C.I.

Accuse not Nature!
She hath done her part;
Do thou but thine!

– Paradise Lost, 8
John Milton

Contents

Foreword

O NLY a small and benighted minority of educated
    human beings now take (or try to take) Genesis
literally. Most of us – even Christians – regard it as a poem,
one of many late Stone Age, or Bronze Age, myths that
have percolated down to us in art and religion. But all such
myths were of course also efforts, if not to explain, at least
to comprehend the external world and man's place in it.
Only fools dismiss them as quaint fossil relics of the past;
one could as well shrug off child psychology as irrelevant
to the study of the adult. We may now have far greater
knowledge of how and when events on our planet have
happened and happen, but we remain still no wiser than
Genesis on why, ultimately, they came and come to pass.
Even the astrophysicists are puzzled by one aspect of the
birth of the cosmos: the extraordinarily narrow limits, in
things like initial velocity of expansion after the 'big bang',
that had to be achieved before a universe that allows our
existence could be.

Robert Fripp's ingenious idea – to resurrect Genesis
in the light of our present knowledge of evolutionary

process – must, I suppose, count as a literary curiosity; but I have long been in favour of literary curiosities. They have a perverse habit, very often, of provoking more thought than the orthodox approach. By simplifying the complex and broadening the narrow, they spark the imagination.

If nothing else, this book should remind the complacent optimists that nothing in evolution, whether divinely inspired or not, suggests that some special place is eternally reserved for our own species. We may think our (comparatively) high intelligence and technology give us a better chance before the kind of catastrophe that ended other non-human creatures in the past; but the problem here is surely that humanity threatens to engineer its own extinction. The past victims of evolution were never menaced by a demon inside their own psyches.

I can just imagine a time when we might quit, or discard, our biological bodies, when the brain might become translated into a machine, immortal, as free of the physical decay of the flesh as of the flesh-caused follies of greed, aggression, selfishness and all the rest. Such a time must lie centuries ahead, yet more and more does contemporary mankind seem to want to forget the biological past, and that Adam and Eve are only one kind of animal among many others, and like them closely dependent on the primordial nature – and limitations – of our planet. Very soon more than half of all humanity will live in cities, and at an increasing remove from biological reality and direct contact with nature. I believe the notion that we are now so sophisticated that we can risk this profound divorce is a sick dream of our century, and shows a monstrous hubris.

The Becoming sharply reminds us that we still remain far nearer the 'lowest' forms of animate life than we do to beings risen, like angels, above their own biology; and that the laws that govern our future remain at heart the same that rule all other existence in a fragile biosphere. Least of all should we ever forget that the god of Genesis also destroyed the cities of the plain.

Let me end with a little personal story. Only a week or two ago I was looking down a microscope at a very rare and recently found local fossil. It was the swimming-paddle, or hand, of a marine dinosaur, the 'fish-lizard' *Ichthyosaurus*. I was looking down a microscope at it because the paddle was of an embryo specimen (this is where its rarity lay), and less than two centimetres long compared to the usual fifty or more. It had been found close beside the fossilized skeleton of the mother from which it had never emerged. I was fascinated, as an amateur palaeontologist; and touched, as a human being. What came, after the first impression of the casual immensity of time, was a sense that it did not exist. This tiny relic of an extinct animal, caught by geological hazard so soon after its own genesis one hundred and fifty million years ago, lay as close to me as my own skin, and equally close inside a common destiny.

Our past remains our now; I pray the reader will not forget that, as he or she goes down the vast vistas of these pages.

JOHN FOWLES
Lyme Regis

Introduction

THE *Oxford English Dictionary* defines the noun 'becoming' as 'a coming to be', or 'a passing into a state'. *The Becoming*, then, expresses the ongoing nature of change and continuing evolution through every age and aspect of our unfolding universe, in a way that the word 'creation' does not.

On the following pages, *The Becoming* presents our contemporary knowledge of evolutionary theories in sixty-two verses written in the English prose style of the King James Version. Each verse, or group of verses, is followed by a commentary in modern English.

§ The intention of *The Becoming* can best be expressed something like this: if the first chapter of the book of Genesis were written today, what would it say? The Creation story would inevitably be different. It might resemble the verses of *The Becoming*.

The human species evolved as an integral part of the natural world. Moments in history when specific human societies decided upon divorce from Nature vary widely:

the oldest city to leave complete ruins, Mohenjo-Daro, thrived for a thousand years, exhausted a forested environment, and died in a desert of its making; goat herds deforested Greece; tropical and temperate rain-forests are being felled in the modern world at astonishing rates. Thinking minds understand that mankind's divorce from Nature must be reversed, and quickly. Bringing this about depends upon achieving the apparently insurmountable task of altering a significant part of modern human mind-set.

But extraordinary changes do happen, and they can happen so fast that the moment of change is unpredictable. Profound change occurs when a critical mass of individuals in society senses that a threat to life and interest posed by change is less than the threat posed by refusing to change.

Much, then, depends upon a society's interpretation of its own best interest. But interpretation is subjective, highly personal. It varies, not only among individuals, but with changing values wrought by changing times. One set of facts is capable of several interpretations, and the fewer the facts in a set, the greater the possible number of interpretations. I give the following by way of example:

An eleventh-century king of England, Canute, ordered his throne set up on a beach at low tide and compelled his counsellors to sit with him while he commanded the tide to stay away. The tide came in, forcing the king and his court to retreat up the beach or be drowned. Those are the *facts*

2

that descend to us through almost a thousand years. Now for the *interpretations*.

Generations of British schoolchildren have grown up knowing nothing about King Canute except that he commanded the tide to retreat, and that it took no notice. The lesson drawn from this is that the man was both arrogant and a fool.

The same set of facts was interpreted differently in the medieval world, which was closer to Canute's time, and still accepted that its roots were of the earth. The medieval view of the incident was this: that the king, in order to impose humility upon arrogant counsellors, forced them to sit with him in the rising waters as long as they could, in order to demonstrate to them the limitations of earthly power.

I suggest that when society can intuitively accept once more the medieval interpretation of the tale of Canute and the tide, it will be better prepared to re-integrate itself into the systems and cycles of Nature. The verses of *The Becoming* are intended as a catalyst, designed to assist open minds to a new (and ancient) view of mankind's place on Earth.

§ After a messy, lengthy and sometimes bloody divorce, science and Christian theology are probing one another's frontiers once again. Like continents drifting apart, science and religion began to seek their own directions during the Renaissance in Europe. The heresy that Earth orbited the Sun did not appear in print until its author, Copernicus, lay safely on his deathbed. Galileo was less fortunate. Confronted by the Inquisition, he trod carefully, dividing the mysteries of the cosmos into two: measurable qualities such as temperature and hardness he defined as primary; subjective qualities such as colour or warmth, secondary. Primary qualities could be safely studied without raising the ire of inquisitors. As to secondary, subjective qualities – among which Galileo was careful not to list religion itself – these fell within the proper purview of the Church. Regarding the intellectual property of priests, he was silent.

René Descartes formalized this rift in terms which came to be called Cartesian dualism: mind distinct from matter, soul from body, God from the cosmos. In this way the new men of science could safely investigate physical manifestations of Creation without seeming to tamper with the spiritual mysteries of a powerful Church. Kepler's work on planetary motion and Newton's on gravity reinforced this dualism, with the curious effect of severing God from the role of Creator. In a medieval age of belief, Thomas Aquinas described God as the 'Prime Mover'. In the new age of reason, God became the clock-maker, the watch winder, the oiler of gears of planetary motion. In such an age, it fell to scientists to reveal glimpses of 'His' purpose.

Dualism persists but, three centuries later, it is science which dominates, leaving a diminished Church groping for relevance. Dualism began as an intellectual device to protect infant science against entrenched religion. The wealth of scientific facts now threatens to eclipse religion altogether. Many scientists feel no reason to invoke the existence of God. Indeed, science's strength in the modern world permits it to sit in judgment as inquisitor upon the Church. Cartesian dualism has turned a half circle.

In keeping with the fast march of science, the story of Creation in the book of Genesis has come to signify an archaic view of origins without relevance or meaning to modern society. Its verses bear no resemblance to the sequence of Creation which the sciences show us, and the majority of thinking people no longer take them literally. In effect, the scope and sense of majestic vision presented in Genesis has been discredited for the modern age.

The mistake of our age has been to dismiss the intent of the work along with the literal interpretation once placed upon it. If we can learn to appreciate the slow, incredible process of continuing Creation, then it may not be too late to re-create on this planet some semblance of that original Eden which was once our inheritance.

It used to be fashionable to explain myth, especially Creation myth, as the proto-science of antiquity, a formula by which the secrets of the physical world were supposedly interpreted in fables suited to an untutored population. In fact myth evolved into other myth, eventually giving rise to the arts and letters; logic and reason were the forces that gave birth to the spirit of scientific enquiry. Between those antithetical castles of fiction and fact lies the great middle ground of imagination and constructive thought; a fertile soil in which the mind's free spirit may set its seed and grow to great height, ultimately influencing everything within its shade. Such is the case with Creation myth, which, in every culture, springs from the compelling human need to search out our collective origin. To know is at least to be certain, even if to be certain is not necessarily to be right.

From which it follows that this is not an attempt to reach a compromise between the literal interpretation of Genesis' Creation on the one hand, and evolutionary theory on the other. It is an attempt, in allegorical form, to combine the spirit and the sense of Genesis with the conventional wisdom of current scientific thought.

ROBERT FRIPP
Toronto, Ontario
Shillingstone, Dorset

6

The Becoming

1. In the beginning was darkness and the silence of the void. And the spirit of God looked out upon the void, and was alone within it.

2. And God stretched forth his hand and gathered about him the ether and dust unto one place. And God took of the ether and kneaded together the stuff of the universe. Then came there forth fires from among it to warm his Creation and shine on his path.

3. So it came to pass that the stuff of the firmament was gathered unto the hand of God. And there came forth rays from among it, that there should be fires, and light. And God said: Let there be light, and there was light.

4. And the spirit of God came forth throughout the whole of it, but the might of his coming was exceeding great, and the fabric of the firmament was rent wide asunder. Then were the ether and dust hurled even to the ends of the heavens, and whirlpools of gas came together in galaxies, spinning white hot in the fires of Creation.

5. And they were consumed utterly. Then God took of the embers of galaxies and moulded the stars and the nebulae of the stars, and God set them among the firmament, even as the multitude of the heavenly host created he them. And God saw light that it was good, and he caused his rays to shine upon all the firmament of the stars, and to shine among them.

6. And God looked upon all the universe that he had wrought, and upon all the firmament of the heavens, and all that therein is, that it was very good.

7. Thus Creation and the dawning were as the morning and the evening of the first day.

8. Then did God move upon the face of the Earth, and saw that it was lifeless and void.

9. Even the mantle of the Earth God found wanting; for God breathed of the ethers that were abroad throughout all the face of the Earth, but among them was the breath of life nowhere to be found, so the mantle was noxious upon it.

10. And God passed over the face of the deep and smote his lightning staff upon the waters that even the ether was subdued and came together unto the lightning which he had wrought.

11. There God forged from the ether the very stuff of life in the womb of the seas; sugars, and amines, and nucleic acids created he them.

12. And God fashioned proteins, and complex large molecules, and to some he gave power of self generation.

13. And God swaddled each one in a cellular membrane, and he blessed them saying: Be fruitful and multiply, each according to his own kind.

14. And it came to pass that the ether and water with the light of the firmament gave some of those succour, and they grew abundantly in the seas that were about the Earth, and gave to her mantle their oxygen. And God saw photosynthesis that it was good.

15. And God said: Let the waters of the seas be filled with life, and let there be green plants and animals in the depths therein, that all shall flourish abundantly and have increase each according to his own kind. And it was so.

16. Thus were the Archaeozoic and the Proterozoic eras even as the morning and the evening of the second day.

17. And the waters brought forth abundantly of moving creatures that have life through every part therein according to the word of God, for the waters gave unto them succour according to his commandment.

18. Life begat divers and myriad forms in the waters, even unto seaweeds that dwell on the rocks of the shore, and sponges that are within the deep.

19. Jellyfish also God caused to swim in the seas that they should be brought whithersoever the wind and the tide willeth, and to have habitation with the plankton both green plants and animals that are within the deep.

20. Then God caused worms to swim in the seas and annelid worms to crawl in the sands thereof.

21. And from some he made molluscs, and gave unto each one a shell and a foot to hold fast as the limpet the rocks of the shore, that these should become as the snails, even as the squid of the seas, and as clams sent he them forth.

22. To others he gave mantles of chitin, and jointed legs paired two by two, that of their offspring should come forth all manner of insects and spiders, and those which swim in the sea, and those which crawl beneath it.

23. And God saw life that it was good, and fashioned each as he desired, the starfish in the depths, and crustaceans that swim in the face of the deep.

24. And it came to pass that seas covered much of the Earth, and the waters brought forth abundantly of coral within the deep. And there came up great pillars of fire upon the land, and hills of smoke from out of the waters that are about the Earth.

25. And God spake, saying: Every valley shall be exalted, and every mountain and hill made low. So God brake the land with wind and rain and caused the dust of the earth to return to the rivers and to enter once more within the deep. Then God caused hills to come forth from the waters and great mountains to come up also out of the seas, that the testimony of his might was graven on dry land, even as the shells that are fossil within the rocks thereof.

26. And God moulded of clay the first creatures with backbones and set them apart, that their issue should come forth abundantly through all the estuaries of the seas, and begat of their offspring the earliest fishes, like unto lampreys that want even jaws.

27. And God breathed of the mantle of Earth that it had oxygen among it, and he sent up scorpions from out of the deep to have habitation upon it, that they should be as a sign unto all that the dry land was good.

28. Thus it was that the Lower Palaeozoic era came to be as the morning and the evening of the third day.

29. And there came forth fishes in the estuaries of the seas, and some begat jaws and became as the sharks, others even as the bony fishes that are within the deep.

30. And those there were which dwelt in rivers parched by the sun, and to these God gave lungs and stout fins of flesh that they might find pools to sustain them with life.

31. Thus crept they on dry land with primitive limbs, even as lungfish which dwell on the rivers' edge.

32. And there were fishes abundantly in great waters. And there came of their issue amphibian, which are as one upon dry land and in the rivers which are upon the land. And they multiplied in the sanctuary of the waters, and went forth upon the Earth.

33. And God made of their substance the frogs, even salamanders made he to be heirs unto them.

34. And God caused plants to come forth from the earth. So there came up from the ground both horsetail plants and the seed-ferns unto great height. And they breathed of the mantle and purified it.

35. Then God gave insects to be among the green plants, even the cockroach and the dragonfly brought he them forth from out of the earth.

36. So God gave unto the dry land life, and caused all manner of creatures to be upon it, and made of the dust both the plant yielding seed, and the plant that hath spores.

37. Thus were the Devonian and the Carboniferous times like unto the morning and the evening of the fourth day.

38. And it came to pass that God set reptiles on the Earth that they should have dominion over it in their time. Even as the dinosaurs sent he them forth in many and divers forms both great and small upon the land, and those there were which ran upon two legs, and those upon four even as the beast of the field.

39. Some sent he to rivers and the swampy ground, and yet others unto the depths of the sea as the ichthyosaurs, that they should return to the place whence life cometh and be abroad among the deep.

40. And God made great turtles to be abroad in the face of the waters, and crocodiles to be about the water's edge: and they were as one with leviathan, which delves in the earth and buries her eggs in the sand of the shore.

41. And it came to pass that God brought forth subtile serpents from out of the earth, that henceforth they should crawl on their bellies and eat of the dust all the days of their lives. So it was that God caused the serpents to go forth as every other creeping thing that creepeth upon the earth.

42. And to some he gave feathers and set them apart as the fowl of the air that they might fly in the open firmament of the heavens; even as the birds of the air sent he them forth.

43. And God made of the reptile the beasts of the field, and gave unto each a coat of hair, and they were as mammals which give suck to the fruit of their womb each after his kind, that they in their turn should inherit the Earth, even to the uttermost parts thereof. And God created great whales from among them, that they should be their sign upon the deep.

44. Of the dust God made cycad trees like unto palms, and the cone bearing tree he wrought also, whose seed is naked unto itself in the earth.

45. Then he created the flowering plant, saying: Let the earth bring forth grass and herb yielding seed, and the fruit tree yielding fruit after his kind, whose seed is in itself upon the earth. Even the fig and the willow tree brought he them forth.

46. Thus were the Mesozoic and Tertiary times even as the morning and the evening of the fifth day.

47. On the sixth day God looked out upon the Earth and saw all life that it was good.

48. And God blessed them, saying: Be fruitful and multiply, and fill the waters of the seas, and let fowl multiply in the air, and let the beast of the field and the creeping thing multiply each according to his kind. And it was so.

49. And God said: Let us make man in our spirit after our likeness; and he brought forth man upon the Earth and gave unto him a part of the spirit of God in his likeness that man should come to have wisdom in all things. For the thoughts of wisdom are more than the sea, and her counsels profounder than the greatest deep.

50. So God created man and gave unto him a part of the spirit of God that in wisdom he should have dominion over the fish of the sea and over the fowl of the air, and over cattle, and over all the Earth and every living thing that creepeth upon the earth.

51. And God blessed them and said unto them: Be fruitful and multiply and replenish the Earth. And God said: Behold I have given you every herb bearing seed which is upon the face of the Earth, and every tree, in which is the fruit of a tree yielding seed, to you it shall be for meat.

52. And to every beast of the earth, and to every fowl of the air, and to every thing that creepeth upon the earth, wherein there is life, I have given every green herb for meat. And it was so.

53. And God saw every thing that he had made, and behold it was very good.

54. And God said unto man: Ask now the beasts, and they shall teach thee; and the fowls of the air, and they shall tell thee.

55. Or speak to the Earth, and it shall instruct thee; and the fishes of the sea shall declare unto thee.

56. Who knoweth not in all these that the hand of the Lord hath wrought this? In whose hand is the soul of every living thing, and the breath of all mankind.

57. But man heeded not the word of God that in wisdom he should have dominion over the Earth, but rather subdued it that it should be according to his will.

58. So man set himself over the spirit of God, even above the spirit of Creation which had brought him forth upon the Earth.

59. But the heavens and the Earth were finished, and all the host of them, and the morning and the evening were as the sixth day.

60. And on the seventh day God ended his work which he had made.

61. And he gave his Creation into the hand of man, that man should be as a steward unto all that the Lord God had made.

62. And on the seventh day God rested, and waited, that he should know how it would come to pass. So it was that man came to have power over Creation, either to magnify or to destroy.

Verse 1

1. In the beginning was darkness and the
silence of the void. And the spirit of God
looked out upon the void, and was alone
within it.

For the authors of Genesis the moment of Creation was
the start of time, the point at which Divine Will began
to create and ordain the destiny of universe. Order was
imposed on chaos and lights appeared in the eternal night.
Contemporary science recognizes a similar moment from
which the universe emerged to flow onward in time, and
out, ever out into space.

Genesis shares the notion of a spontaneous beginning
with the Big Bang theory, widely accepted by many scientists as explaining the origin of the universe. Whether one's
belief is based upon faith or the evidence of the stars it is
difficult to conceive of anything before that unique instant
of beginning, except, of course, for darkness and the
silence of the void. Scientifically, our universe appears to
be an effect for which we hope to find a cause, but no
amount of investigation has so far revealed the primary
force by which it was created. It has been suggested that a

previous entity collapsed into itself, giving birth to our ancestral fireball, but that is no solution to the initial problem. Anyway, the evidence for such an event would have been destroyed in the instant of our own Creation. It may be that the remnants of a different space and time escaped the conflagration, lingering still on the outer fringes of our cosmos, but that we cannot know.

We deceive ourselves into believing that if we could penetrate the farthest limits of space/time we could return with an answer to resolve the difference between the Western tradition of a single and unique Creation and the recurring cosmic cycles fundamental to Eastern spirituality. Even then we should have explained only the effect, and not its cause.

'In the beginning God created the heaven and the earth.' We are culturally conditioned to regress to 'In the beginning ...' Better, surely, to imagine the presumed moment of 'beginning' as the Instant of Realization. In terms of a governing cosmic force or intellect, that is what it must have been. Shifting the emphasis allows our limited comprehension to ponder that colossal moment without falling into: what existed before the beginning? before there was anything at all?

If we could only know the primary cause ... But then, if the elusive secrets of the cosmos stood revealed, we could no longer seduce the beauty of its mysteries. The author of Ecclesiasticus wrote of such wisdom: 'The first man knew her not perfectly: no more shall the last find her out.' After all, if one were all-knowing and all-wise one would be God. If there is one.

If God does not exist, the classical Chinese model of the cosmos will serve as well as any. There, the universe becomes an endless 'web without a weaver'. In the Western sense the very word 'Creation' implies that it is the product of a higher intellect, which in turn elicits images of primordial chaos brought to order, substance and form by the conscious act of a Creator. Eastern traditions do not view the cosmos as the manifestation of a single God. Rather, they hold that it exists because it exists, uncreated, unending, forever changing in order to remain the same, and driven by the dictates of its own dynamic nature. Western man prefers to fall back on the comforting notion of Creator. Thus Emerson: 'There is never a beginning, there is never an end, to the inexplicable continuity of this web of God.' Emerson seems to want to impose a Western-style God on the *Tao*.

To the Eastern mind, the eternal process unfolds as a balance between two opposed or complementary forces. As such, one might view cosmic evolution as an unceasing contest between two polar spiritual complements: *Yin/ Yang*; *Yab/Yum* (father and mother, in Tibet); Good and Evil; Darkness and Light; God and the Devil in some form or another. And then there is Shelley:

> To suppose that the world was created and is superintended
> by two spirits of a balanced power and opposite dispositions,

is simply a personification of the struggle which we perceive in the operations of external things as they affect us, between good and evil.

Even if one denies an opposed duality in the spiritual world, one recognizes its physical counterparts: order and chaos; matter and energy; space and time; day and night; heat and cold. Gravity holds galaxies together while centrifugal forces throw them apart. And stars would collapse beneath their weight were it not for nuclear forces supporting them. Thus nature appears in constant conflict with itself, its energies for ever trying to cancel one another out.

But where one would expect the resulting product to be nothing, it becomes, well, everything. Such are the apparent contradictions by which complementary forces combine to form the perfect whole.

It is easier to comprehend this state of affairs from a Taoist or Vedic point of view than from a Judaeo-Christian one. True Buddhist enlightenment reflects the emancipation of a mind that has learned to reconcile Creation's apparently opposing natures. On one hand Creation exists through its dynamic, ever-changing material forms; on the other hand it manifests itself as passive wisdom, constant, absolute and undefilable, transcending all apparent contradiction. The Buddha taught that such a reconciliation must be experienced through the enlightenment achieved by meditation. Enlightenment can never be gained through intellect alone. It stems from the insight that all apparent opposites are manifestations of a single, indivisible whole. Buddha remarked of his own enlightenment: '… Ignorance was dispelled, and science [true knowledge] arose; the darkness was dispelled, and the light arose.'

Western precepts prefer to define the nature of Creation rather more precisely. Our materialist perception demands a well-defined Creator, one that will eventually triumph over the dark satanic forces of our insecurities. Surely, we like to think, only God's light could penetrate the darkness of the original void, and only God could frame order out of chaos. 'He had First Matter seen undrest … before one rag of form was on.' * The materialist tradition in Western culture goes back to the first known philosophers, descendants of Animists for whom external forms of matter were the merest accident. God, or Manitou, might speak as clearly from a rock, a hill, a burning bush or the freshly-killed carcass of a hunter's caribou. An unwitting slave to his ancestral precepts, even Thomas Aquinas found it necessary to fall back upon physical analogy to demonstrate the existence of God. All things, he wrote, are moved by other things, smaller ones by larger, and so on up the line. From which he inferred that every thing is ultimately moved at the whim of a Prime Mover, which he called God. Aquinas found the idea of a self-directing universe inconceivable – as absurd as the notion

* *Hudibras* I.i.562, Samuel Butler

that a hammer and saw should build a box without a
carpenter to guide them. Aquinas saw the cosmos as the
product of the first, the uncaused, cause – the carpenter –
'which we call God'. As such we can define God, assigning
attributes to It which fall within the framework and pur-
poses of human understanding. Thus, God is good, or to
be loved, or feared. But, as Aquinas put it, we can never
know directly what It is.

On that last point, Aquinas seems in agreement with
Hindu cosmogony. Brahman, the 'Absolute', may choose
to make Itself manifest. Thus we are told, 'All that exists is
Brahman.' At other times, 'Brahman is truth.' By its very
nature, Brahman, like Thomas Aquinas' sense of God, is
impossible to confine, let alone depict.

Another notion of Creation is more sentient, more
human. Hardly surprising, then, that it falls between the
concept of a godless eternity and one in which God is
revealed through the physical dynamics of Its universe.
J. B. S. Haldane makes the case for a universe that is much,
much more than one of blind mechanism. It becomes for
him a spiritual entity which, though seen imperfectly, has
neither matter nor motion as its central truth, but rather
the force of mind. Having gone that far, Haldane ventures
to suggest that the cosmos might not only be queerer than
we suppose, but queerer than we can suppose. That
accords well with Emily Dickinson's contention: 'Had we
the first intimation of the Definition of Life, the calmest of
us would be Lunatics!'

Better to set insanity aside and take up reason. The physical sciences yield valuable indirect knowledge of an emerging concept of universe, but, even so, such knowledge can never take the place of direct perception. For example, a blind person can infer a great deal about sunlight from its warmth upon skin, going on to make accurate inferences about the nature of its source. Inference, however, is a poor substitute for sight. We are all similarly limited by physical senses that cannot adequately assess the nature of consciousness, the major tool we use to form our own ideas of universe.

That sort of awareness will not come home to us just because we look ever farther outward with the aid of telescopes or sophisticated inter-stellar probes. It will come when science turns itself around and looks inward with a perceptive inner eye to find that God, reality and the universe are really very close as well as far away. God may be no more than consciousness, and those who, in a material age, can still contemplate the night sky and wonder at the mysteries of Earth will find in themselves the cosmos as well as its microcosm. When we understand that, we shall be able to straddle the universe with perfect ease, and measure its dimensions.

Verses 2 and 3

2. And God stretched forth his hand and gathered about him the ether and dust unto one place. And God took of the ether and kneaded together the stuff of the universe. Then came there forth fires from among it to warm his Creation and shine on his path.

3. So it came to pass that the stuff of the firmament was gathered unto the hand of God. And there came forth rays from among it, that there should be fires, and light. And God said: Let there be light, and there was light.

LET'S start at the beginning, at our arbitrary concept of a beginning, the Instant of Realization.

Edwin Hubble recognized in the 1920s that the universe is expanding, each part of it receding from every other part as its envelope grows. To explain this, Abbé Georges Lemaître offered what has come to be known as the Big Bang theory. Lemaître, professor of relativity studies at Louvain University, was an astrophysicist-priest who spent the latter part of his life attempting to reconcile the Genesis account with what was known of cosmic evolution. He suggested that the universe had developed from a kind of 'cosmic egg' into which all matter had been sucked, and compressed, by the force of its own gravitational pull. Recent theory goes beyond Lemaître's cosmic egg, suggesting that, some fifteen billion years ago, the whole enormity of Creation derived from a single point. Theory goes on to describe that once-and-forever point as being greatly smaller than a proton, which is itself a sub-component of an atom in our given scheme of things. (If an atom were the size of a football field, the proton at its core would be smaller than the ball.)

There is something of the grotesque, a genie-in-a-bottle quality, to this theory that everything in the cosmos emerged in an instant and continues to flow onward and outward, emanating from a point which we represent to be virtually nothing.

Everything from nothing. To imagine such a feat comes as something of a shock, yet that is precisely what two major religious traditions have suggested all along.

In the biblical account, God 'created the heaven and the earth'. From the darkness and the void, presumably. The author of Genesis was never troubled by a sense of What Came Before? His God was assuredly that of Thomas Aquinas – the Prime Mover – constructing a physical world from first principles. The materialist assumption of a definite starting point was taken for granted.

Similarly, in the Vedic tradition, the primary belief is of

an uncreated intelligence, eternal in time, indefinite in space, comprising all that is 'being and non-being': that is the only reality, the ultimate cause and source and purpose of all that exists. This is Brahman, the Atman, the Absolute, the All. In contrast to Genesis, the Absolute causes Its universe to emanate from itself and to reabsorb itself, continually transforming its energies and physical forms. For three millennia, devotees in Asian traditions have sought to discover the same goal – through meditation, as Buddha instructed – that modern scientists are seeking through intellectual pursuit. They search for the Ultimate Reality – the All.

Science and religion may be working toward the same Absolute Truth. But for those brought up in the biblical tradition it is easier to comprehend the religious notion of 'conceptual nothingness' at the point of a definite 'Time Zero' beginning, rather than the 'spatial nothingness' suggested by science. This is not surprising. We can grasp the conceptual nothingness of non-existence in non-time. Such was our state before our parents gave us life. It is harder to accept that cosmic evolution sprang from the spatial nothingness dictated by mathematical theory. Leaving science aside, the notion seems to violate every rational principle learned since childhood. As infants we discover that large pegs do not fit into small holes; as adults we have difficulty parking large cars in small spaces. We are born to experience the solidness of things. Every concept of spatial relationship rebels at the notion that the entirety of the universe once flowed from a single point, the dimension of which was a virtual nothing.

But why not? One might as well ask: what size is a thought? It has no size, no mass, no rest-energy, as a physicist might say. And yet the power of directed thought is the greatest single force in binding, dividing, creating and destroying that fragment of consciousness we know as the human condition.

Now extend the parameters of this argument to include a greater consciousness and a greater thought. Let it become infinite, eternal; for ultimately, if one believes in a God/Creator, then the true nature of Creation and its universe stands here clearly revealed. The whole of Creation is no more and no less than an ancient, ever-growing thought.

In an instant [one] poynt that is [produced] fillyth all the world of light and shining.

– Bartholomeus Anglicus,
translated by John de Trevisa, 1398

§ Let's assume that it happened that way. Let the universe be condensed to a point several billion times smaller than a proton. Whatever exists does so as pure, seething, unrealized energy, hot and compressed beyond comprehension. Matter cannot exist in any form, except perhaps as a concept of the Absolute and as a destiny. The theory is easier to absorb if one can rationalize it. Imagine that this instant of primordial energy is the concept of universe – the thought, not the universe itself – which is waiting, straining to emerge and to be.

At this stage, theory says, all natural forces – gravity, strong and weak particle interaction forces, electromagnetic energy – are subsumed into one entity, a unified force, rich in potential. The four forces of the cosmic apocalypse have yet to emerge. Such is the situation at the beginning.

Let it happen. Now!

The Instant of Realization. The Big Bang. Time Zero.

Even theory is at a loss to explain the very first events in cosmic evolution, but it soon finds its way, at 1×10^{-43} part of a second after the Instant of Realization, to be precise. The expanding universe has attained a diameter of 1×10^{-27} of a millimetre. With a temperature of 10^{32} degrees Kelvin, the infant fireball is inconceivably hot, but cooling rapidly as it expands.

Now gravity becomes its own force, no longer subject to the previously unified and universal power binding all the energies together. Here's an irony: the moment at which the Absolute imposed Its Will upon the nature of Creation was the very moment in which that same Creation passed beyond its own age of absolutes. Henceforth, according to a principle of general relativity, gravity will exert its influence through all eternity by imposing a relativistic warp on the whole fabric of space and time.

At 1×10^{-35} part of a second after the Bang, the model of the expanding universe enters something called its 'inflationary' phase. An 'Inflationary Universe' is the intellectual property of Alan Guth, who calculates that space went through a rapid expansion at the moment when the once-unified force began to differentiate. In this instant the independent strength of gravity threatens the growth of a naissant universe (and the Big Bang theory), so it is essential that a force opposed to gravity should exist as a counterweight. It may still exist, diluted beyond measure.

Anti-gravity serves several functions in this scheme of things. In the first place it suppresses the coagulation of massive particles that would otherwise be formed by the first-born stuff of matter, quarks and electrons. Anti-gravity also acts to repel all things in the miniature cosmos evenly, causing its fireball to grow more rapidly than before. This force prevails while the new universe attains the size of an orange, some eight to twelve centimetres (three to four inches) across. Without an anti-gravity force early on, one theoretical model predicts the collapse and implosion of the embryonic universe after little more than thirty thousand years.

Now energy begins to congeal, losing its purity and degenerating into the baseness of matter, as if it were water vapour freezing into ice fronds on cold glass. Quarks, the building blocks of protons and neutrons, are created here; so are negatively charged electrons and their counterparts in anti-matter, positively charged positrons.

Near the end of this rapid-expansion phase, strong and weak particle interaction forces emerge to take their present forms. The universe has been cooling as it expands, but the liberation of the strong force, which binds protons and neutrons in atomic nuclei, temporarily heats it up again.

At Big Bang plus 1×10^{-32} of a second, the universe leaves this inflationary phase behind. Its temperature drops to 10^{27} degrees Kelvin. The cosmos is now a grapefruit-sized body of uniform density in which matter exists, though not in its final form, side by side with anti-matter. Once more a harmonic symmetry reasserts itself, briefly, for anti-matter is identical with matter, except that it carries an equal but opposite electrical charge.

Now, for a short time, we enter a zone of numbers we might be able to grasp. After the first millionth part of a second, the universe has grown to about the size of our solar system, its temperature dropping to around ten million million (10^{13}) degrees Kelvin. At this lower temperature, quarks bind together to form protons and neutrons. The cement for a material universe is beginning to fall into place; but then, in this instant, matter and anti-matter annihilate each other and the illusion of material symmetry is lost. A slight surplus of *matter* survives, as it must, to form the stuff of stardust, the material of the universe.*

Big Bang plus one second, precisely. Wraith-like particles called neutrinos come into being. Strange entities, they seem able to break free of bonds constraining other

---

* A 1996 experiment at CERN (no. PS-200) showed that matter does not anihilate antiprotons as fast as cosmic theory would suppose.

matter. In some ways they resemble electrons except that they have virtually no mass, and unlike electrons they seldom interact with other particles because they carry no electrical charge. However, they clearly have a purpose in the one second old Creation, because the great flood of neutrinos created in the Big Bang accounts for perhaps nine tenths of the universe's total mass. Despite which, neutrinos are the most independent of all known sub-atomic particles, shunning attachment to others and to the physical fabric of the cosmos. And yet ...

'A web without a weaver' is how Chinese tradition perceives the universe. Give that web three dimensions and it begins to resemble the fibrous skeleton of a sponge, in which a lattice of filaments encloses large gaps of nothingness.

Galaxies and intergalactic dust-clouds are distributed in ways that resemble the fibrous filaments bounding those pores in the sponge. We find them in clusters or strewn out in well-defined strands that form the boundaries of voids hundreds of light-years across. What we are seeing in our present cosmos resembles the fibrous, sponge-like structure suggested by theoretical models for its early stages. One theory explains this structure by concluding that, after the Big Bang, the first outpouring of neutrinos was not evenly distributed but clumped into well-defined clusters and strands. Assuming neutrinos have mass, the gravitational pull of these neutrino strands would have attracted the first particles of matter as it condensed from

energy. Attracting more and more matter, the gravitational
fields of the strands increased so that they grew by accretion at the expense of the voids between. It looks as though
the cosmic present is constructed on the still expanding
framework of a very distant past.

Neutrinos, then, may fill the role of neutral vessel within which to arrange the principal parts. 'Shape clay into a
vessel,' says the *Tao Te Ching*. 'It is the space within that
makes it useful. Therefore profit comes from what is there;
usefulness from what is not there.'

Big Bang plus three minutes. Temperature drops to
one billion (10^9) degrees Kelvin, cold enough for protons
and neutrons to combine, with the strong particle force
henceforth gluing them together to form atomic nuclei.
The simple nuclei of helium and heavy hydrogen (deut-
erium) materialize at this point, but without orbiting
electron shells. Electrons are still too energetic to be captured by atomic nuclei and locked into orbit. So electrons
continue to wander, unattached.

Half an hour passes. Helium nuclei have captured many
of the heavy particles by this time, taking them out of circu-
lation. As a result, nuclear processes slow down. At this
point electrons and positrons annihilate each other. Just
enough electrons survive the encounter to offset the positively charged protons in atomic nuclei. Unattached protons will become nuclei for hydrogen atoms. But not yet.
The universe remains too hot for atoms to bind together.
And electrons are still free to wander where they will.

A hundred thousand years into Creation's scheme of things the temperature drops to 3,000 degrees Kelvin. As the new universe cools, electrons lose enough energy to be captured by atomic nuclei and locked into orbit around them. With protons and neutrons for atomic kernels, and electrons for atomic shells, matter has found its form for all time as fully-fledged atoms. The enslavement of energy into matter is well under way.

As mud settling to the bottom of a pond leaves the water clear, so the capture of electrons into atoms leaves great spaces in the universe through which radiation can shine. 'Lighten our darkness, we beseech thee, O Lord!' That moment is now. Photons – energetic particles of visible light and other electromagnetic forces – come into their own. The dull glow that suffused the impermeable fog of the youthful universe begins to change, yielding in time to its present aspect, with pinpoints of starlight gleaming through a clear, pellucid dark.

It took about a hundred thousand years for that which is strange to yield to something approximating the universe with which we are familiar. It would take billions more years before our small corner of the cosmos gave life to our species and endowed it with consciousness enough to ask questions about origins.

§ Anti-matter is back. In January 1996, physicists at the CERN particle physics laboratory, Geneva, announced that they had created nine, possibly eleven, atoms of anti-matter. They achieved this by shooting negatively charged protons (anti-protons) into xenon gas. A few of those particles struck xenon atoms, dislodging positively-charged electrons (anti-electrons). Rarely – nine or eleven times – these anti-electrons were swept into orbit around the anti-protons, creating anti-matter, specifically anti-hydrogen.

This 'Through the Looking Glass' manifestation of God's other hand lasted for one forty-billionth part of a second (1×40^{-9}) from birth to annihilation, at which point matter destroyed it again. If the fury of Creation had gone the other way, if anti-matter had prevailed over matter, would the universe appear exactly the same? That has been the accepted 'folklore' of science until recently. But an anti-matter anti-universe might possess an anti-gravity, a force that repels rather than attracts.* Is it just possible that, 'In the beginning', our universe briefly took its explosive expansionary force from anti-gravity? Such a force would die with anti-matter, but not before it had irrevocably driven each atom away from its fellows into worlds with-out end, Amen.

§ A note about the apparent nonsense of times like 1×10^{-35} part of a second:

General relativity theory describes how time slows down as the strength of gravity increases. An atomic clock

---

* Einstein's equivalence principle suggests that gravity should have the same effect on matter and anti-matter.

would run more slowly on the Sun than on Earth. Conversely, it would run faster in the Moon's reduced gravity, and faster still in deep space. In similar fashion, an atom will radiate light at a lower frequency on the Sun than on Earth. However, gravitational disparities between such bodies are comparatively slight, so time and emission-rate differences are correspondingly small.

Albert Einstein's prediction of this effect was eventually proven with the aid of a white dwarf star, the binary companion of the Dog Star, Sirius. The dwarf star's matter is crushed to such density that a handful would weigh several tons on Earth. This enormous gravitational field *does* change the frequency of the dwarf's radiation measurably. Time, on Sirius' companion, known to astronomers as the Pup, runs more slowly, as the 'Einstein Effect' predicts.

Imagine this effect acting through the inconceivable densities and gravitational fields attending the first few seconds and minutes of the universe. If the principles of general relativity were operative in those first fractions of time, one might perhaps fit a large part of eternity into the span of 1×10^{-35} part of a second.

'A thousand ages in thy sight are like a morning gone.' So says the Christian hymn, and in general terms the concept of elastic time is as old as human experience. A medieval German parable makes the point:

In a monastery deep in the forest a monk asked his abbot about tasks the chosen few would perform in heaven through all eternity.

'They meditate and think of God,' said the abbot.

'But, father, how can they possibly meditate so long?'

The abbot made no reply, so the monk returned to work in the fields until he noticed a wondrous bird with plumage the colours of a thousand rainbows. Fascinated, he followed it from tree to tree until evening forced him to abandon the chase. Returning to the monastery he found that the older monks had died and the abbot was now an old man.

'Now you see how it is, my son. If a simple bird turned so many of your years into an afternoon, think how eternity will pass in the company of the chosen few.'

So goes a Christian parable. In Hindu tradition, the birth, life, and death of the cosmos is measured in multiples of a day and a night of Brahma, each lasting over eight and a half billion Earth years ($8 \cdot 64 \times 10^9$).

Disciples of Jainism represent the passage of time as a twelve-spoked wheel revolving once in two million million years (2×10^{12}). Given the extremes of gravity/time compression in which our cosmos was born, it is interesting to speculate how far through its cosmic cycle the Jains' wheel might have turned during what we take to be the first few minutes of Creation.

Verse 4

4. And the spirit of God came forth throughout the whole of it, but the might of his coming was exceeding great, and the fabric of the firmament was rent wide asunder. Then were the ether and dust hurled even to the ends of the heavens, and whirlpools of gas came together in galaxies, spinning white hot in the fires of Creation.

GALAXIES are the basic structures of our universe, island continents of gas and stars, long separated from each other in time and in space. Before their true nature was known, even before astronomers could discriminate between mature galaxies and intergalactic gas clouds (nebulae), Immanuel Kant had suggested that all celestial nebulosities might be 'island universes' in their own right. In a sense they are. Singly, or in clusters, these solitudes of stars and gas have been drifting ever farther from their fellows since the instant in which the new order was born.

Galaxies are so often found in groups, and so seldom found alone in space, that it looks as if clusters, rather than single units, comprise the universe's basic building block.

At their most humble, clusters form simple pairs, like the Whirlpool Galaxy (M51), in which the smaller spiral connects to the larger via trailing streams of gas and stars. Spiral form was first detected in 1850 when the Earl of Rosse discovered the Whirlpool Galaxy. Hence its name. At the time, its discoverer thought that its spiral arms might be exceptional, perhaps unique.

Discovering the Whirlpool's spiral structure proved a boon to astronomers attempting to interpret our own Milky Way. We float through space on one of the largest in a cluster of twenty galaxies known as the Local Group. Clusters like ours are bound by the collective gravitational field of their members. Thus the spiral galaxy of Andromeda (M31) is both the Milky Way's twin in the Local Group, and its counterweight. Andromeda rides through space 2·2 million light-years away from us, half the diameter of the Local Group. We are converging. Five billion years from now the Milky Way will absorb Andromeda, compressing the gases in both. When that happens, millions of new stars will ignite.

If our own ship is humble, others are not. At the other end of the scale many hundreds of separate galactic systems cluster together in such constellations as Virgo and Coma Berenices.

New insights into galactic structure came after 1990, when the Hubble Space Telescope (Hubble) went into an orbit 600 kilometres (370 miles) above the distortion caused by Earth's atmosphere. Images gathered by Hubble have given Earth-bound researchers extraordinary insights into the structures, distribution and dynamics of galaxies and galactic nebulae. Its space-based platform gives the telescope such an advance in resolution over earth-bound instruments that certain classes of astronomical observations fall into pre- or post-Hubble categories.

As Hubble sends back new and astounding galactic images, it is becoming clear that Kant's notion of 'island universes' describes their distribution in the present rather than in the distant past. Hubble has captured images from an ancient past when galaxies collided, merged, tugged their neighbours out of shape and fed on each other's gas. Hubble's pictures of objects eleven to twelve billion light-years distant and as many years ancient give us random snapshots from a slow-motion war of the worlds when galaxies penetrated and cannibalized each other or sucked another's star-stuff into their black holes.

That is another proof that Hubble's photographs bring home to us: black holes are everywhere. The Space Telescope Science Institute's Mario Livio comments: 'There is a massive black hole' in the core of every galaxy that Hubble has studied.* A black hole is a zone of such crushing density and gravitational attraction that even light is trapped. The black hole phenomenon appears to reverse the sequence from which the universe began: a black hole

---

* *Newsweek*, 3 November 1997, p. 33.

attempts to suck the genie back into its bottle, as it were. Our Animist ancestors understood Creation as a continuing process, an eternal rebirthing of substance and form. If Big Bang Plus represents an ancient, ever-growing thought, perhaps a black hole represents a lesser cosmic thought refocusing its energies in preparation for an explosive new coming-to-be. Where does a black hole lead? To another crush of matter the size of a proton? And what will hatch out? A new nebula, perhaps.

How did galaxies form? A short answer is that each is the product of a proto-cloud of helium and hydrogen collapsing into itself, falling victim to its own gravitational field. From that point on the process has been reconstructed by studying the workings of our own galaxy, the spiral Milky Way.

Imagine the initial collapse of a gas cloud falling in on itself, compressing into a ball of plasma that spins ever faster on its axis as it shrinks, becoming hotter and more dense. These early stages of collapse may have given birth to the first stars, their fusion-fires intense enough to forge some of the heavier elements in their cores, including metals.

The fate of the galactic cloud itself provides the major drama. It becomes a creature in the throes of its own Creation. Eventually it reaches a point at which its internal stresses are too great. Henceforth the crushing power of its gravitational field is opposed by the outward thrust of nuclear forces and the centrifugal effect of rotation.

A sudden eruption ejects two plumes of superheated gas, one on either side of the fireball: action, reaction. Expelled from the central cloud, these streams of white-hot plasma trail behind, forming spiral arms as the hub continues to spin. A vast, celestial Catherine-wheel offers a useful analogy for a spiral galaxy's function and form.

Nothing is certain. Many astronomers were confident that spiral galaxies like ours constituted the 'pure' form, the archetype, the first to evolve. Where two or more spiral galaxies collided, theory supposed, the merger resulted in the next most prevalent form, a crude elliptical shape. Hubble findings may correct this view: elliptical galaxies may be older.

In December 1995, Hubble Space Telescope director Bob Williams pointed the instrument for ten days at one tiny patch of sky. (A pinhole in a piece of paper held at arm's length gives an idea of the sky-field involved.) The resulting image, called the Hubble Deep Field, reveals the most detailed image to date of the distant – and ancient – universe. Hundreds of galaxies in all sorts of shapes shine at us across billions of light-years through this 'pinhole' of sky. These images are astounding not because of the distance involved, but what that distance represents. Hubble is showing us galaxies as they looked between 10·4 and 11·7 billion years ago, when the youthful universe was smaller and its galaxies newly formed – formed moreover of hydrogen and helium, pristine material from the cosmic fireball. The Hubble image reveals: that galaxies were more numerous in the past; that many thousands of stars ignited more or less simultaneously (the modern figure for our own galaxy is just two or three a year); and, that about one third of the galaxies visible in that ancient time were interacting with each other, colliding, draining or distorting one another.

Hubble Deep Field has retrieved for us a time *before* galactic clouds had been adulterated by heavier elements: such elements had yet to be forged in the fusion fires of unborn stars. The image is a snapshot of the universe from an ancient time before the planet Earth and our sun were born.

Other Hubble images show what astronomer Rogier Windhorst calls 'subgalactic clumps … eighteen little blobs, eleven billion light-years from Earth'. Each blob contains about one billion stars. To be accurate we should say 'used to contain', because Windhorst is seeing the clumps as they were eleven billion years ago. What may once have been building blocks for galaxies have since grown, merged, perhaps self-destructed or been sucked through black holes to burst forth as visible matter again. Williams' images and Windhorst's blobs represent the baby pictures of Eternity.

Ancient, and modern. The cosmos renews itself, too. Creation is an ongoing process, a continual *becoming*. In 1993, a team led by Howard Yee claimed to have found the youngest galaxy so far discovered, ten million years old, barely a tick of the cosmic clock. One hundred times brighter than our Milky Way, Galaxy MS1512-CB58 owes its brightness to youthful exuberance: it is giving birth to the equivalent of one thousand suns a year.

What holds such vast structures in cohesive shapes? Both gravitational and electromagnetic forces have been suggested as binding mechanisms. Imagine the spreading pattern of concentric ripples when a stone strikes a pond. If, in addition, the water at the point of impact were revolving like the vortex of a whirlpool, the result would be a series of waves moving away from the centre, not as concentric rings, but as spirals. One elegant theory proposes that density waves may move out from the galactic hub in that manner. A density wave effect would help maintain the grand design of the spiral pattern as well as explain the puzzle of star formation in the spiral arms.

Galaxies' spiral arms are active star producers, despite having a low gas density. But it takes a high gas density to trigger the kind of gravitational cloud-collapse that gives birth to a star. Why else would a gas cloud collapse? And how to explain this contradiction? Outward flowing density waves – the pond ripples – would compress the gas in the arms through which they pass, not for long in astronomical terms, but long enough to bring gas clouds to a state of collapse. Stars would thus form almost simultaneously along a broad front in the spiral arms. Dust lanes occur in our own galaxy on the spiral arms' inner edges. Several light-years farther out – farther from the galactic hub – the bright glow of young stars shines through the gas of the spiral arms proper. This apparent discrepancy may represent the time and distance necessary for the outward-flowing material to collapse and coalesce as stars.

In 1995, Hubble captured a superb example of a shock/density wave effect. Eons ago, a small galactic nebula drove its way through the hub of a larger spiral, the Cartwheel Galaxy. The impact's shock waves destroyed the Cartwheel's spiral arms and pushed gas clouds outward from its hub until they reached sufficient density to condense and fuse into a perfect circle of stars having a radius tens of light-years distant from the hub.

In the early 1940s, Walter Baade used the Mount Wilson telescope to study the Andromeda nebula. Andromeda's main spiral is attended by two smaller satellites, the elliptical galaxies NGC 205 and NGC 221. Baade's major contribution to astronomy was to show that stars in elliptical galaxies and the hubs of spirals are different from those found in spiral galaxies' trailing arms.

Stars in the hub, Baade's 'Population II', tend to be small. Their light appears red or yellow because they are not large or dense enough to burn brightly. Such stars exist in a space free of dust and gas, and are very long-lived. Their hydrogen and helium stores may fuel them for billions of years.

By contrast, at the other end of the stellar spectrum, giant stars in spiral galaxies' arms burn blue-white. Baade named these 'Population I' stars, noting that they associate with dark inter-stellar dust clouds or bright nebulous material. These blue giants dominate the spiral arms, but they burn so brightly, emitting radiation at such a rapid rate, that they consume their hydrogen fuel very fast. Their lives may be measured in mere tens of millions of years, but they are so abundant in the spiral arms that they must be continually forming, dying, and forming again.

In life, blue giants' internal fires are so intense that in death they leave a legacy of heavy elements to be taken up by embryonic stars of succeeding generations; and when youthful stars rise like phoenixes from their predecessors' ashes, their own fusion fires serve again to increase the proportion of heavy elements in their cores by forging together the atoms of lighter ones.

That much has been charted, for the oldest stars – the reds, formed from primordial clouds of helium and hydrogen – have but a tiny proportion of heavy elements in their composition. By contrast, young, hot, blue stars, entrapped in clouds and bright gaseous nebulae, show metals and heavy elements in their spectral lines.

There is a progression here: the hub is a turbulent gas-cloud in which the oldest stars coalesce from the lightest elements; but, in the spiral arms, successive generations of short-lived stars create and comprise ever greater proportions of heavy elements in their composition.

It is humbling to recall that the chemical elements making up our bodies and our planet were forged in the fusion-fires of stars. The carbon of our flesh and the calcium of our bones are not the end-products of just one star's birth, life and demise. Star after star has ignited, burned and vanished into gas and ash. Over time, the dust of one has mixed with fuel for the next, each cycle leaving a higher proportion of heavy elements behind. Eon on eon, star after star. We descend from a very long line of long-since born and long dead stars.

Verses 5, 6 and 7

5. And they were consumed utterly. Then God took of the embers of galaxies and moulded the stars and the nebulae of the stars, and God set them among the firmament, even as the multitude of the heavenly host created he them. And God saw light that it was good, and he caused his rays to shine upon all the firmament of the stars, and to shine among them.

6. And God looked upon all the universe that he had wrought, and upon all the firmament of the heavens, and all that therein is, that it was very good.

7. Thus Creation and the dawning were as the morning and the evening of the first day.

THE cosmos is no static relic of an ancient time. It grows and changes, an ongoing *becoming*, like the living thing it is. New stars are born from the gas of inter-stellar space and the dust of their forebears; others die, either dwindling to the impotence of dark dwarf stars, or exploding in a posthumous blaze of glory, their passing marked by clouds of debris light-years across. Many such clouds must once have been magnificent, for they represent the long-faded supernovae of exploded stars. Such spectacular epitaphs. Of many a dying star it might be said that nothing in its life became it like the leaving it.

Few humans have the luck to witness the brilliant splendour of a supernova: the Crab is one of few recorded in the past thousand years. The bright flare of the star's demise reached Earth in July 1054, to be recorded by Chinese court astronomers, among others, who reported a 'guest star' more brilliant than anything ever seen in the night sky. For months, nights were bright enough to read by, and the 'guest' was visible by day for almost two years. Reduced to a shadow of its former glory, the Crab's debris is still scattering, some five thousand light-years away.

Within the span of a human life the night sky seems changeless, but the familiar shapes of constellations are very different from the outlines known to navigators and shepherds of biblical times. How bright the night sky must have been when the Earth was young and the universe itself was younger. Its light was then contained within a smaller span. Perhaps on the juvenile Earth the nights were rival to the day in brilliance.

§ It may be a fool's task to try to preserve an abstract sense of wonder in the modern age, for science has unravelled much of the mystery and most of the magic. However, the correlation between the human spirit's capacity for wonder, on one hand, and the light of science and reason, on the other, is a little like the relationship between energy and matter. In both cases the former gave birth to the latter. So, to help maintain a sense of wonder in a summary of these first seven verses, it is best to go back and ponder the established order of the universe through a mind whose eighteenth-century perspective combines the instincts of both ancient and modern times.

Joseph Addison was an English poet, politician and diplomat. He is best remembered for several hundred essays published in the *Tatler* and the *Spectator* from 1709 to 1714. It was Addison's good fortune to live in an age when English letters flourished and were well received. And it is our good fortune that he was a true product of his day, a man who is modern, and yet not quite modern.

A full century and a half had elapsed since Nicholas Copernicus' *De Revolutionibus Orbium Coelestium* had been published, its author lying safely on his deathbed. Copernicus lived with his knowledge of a sun-centred solar system for years, withholding publication for fear of offending the Church. Four centuries later, another priest, Georges Lemaître, would freely propose that the universe began as a 'cosmic egg'. The Big Bang theory and the Inflationary Universe descend from Lemaître's work as products of mathematical reason. Somewhere between religion and reason lie the thoughts of Joseph Addison, son of a clergyman, whose essays managed to combine reason in the modern age with an older sense of reasoning belief:

I was yesterday about sun-set walking in the open fields, until the night insensibly fell upon me. I at first amused myself with all the richness and variety of colours which appeared in the western parts of heaven; in proportion as they faded away and went out, several stars and planets appeared one after another, until the whole firmament was in a glow. The blueness of the ether was exceedingly heightened and enlivened by the season of the year, and by the rays of all those luminaries that passed through it. The galaxy appeared in its most beautiful white. To complete the scene, the full moon rose at length in that clouded majesty which Milton takes notice of, and opened to the eye a new picture of nature, which was more finely shaded and disposed among softer lights than that which the sun had before discovered to us.

As I was surveying the moon walking in her brightness, and taking her progress among the constellations, a thought rose in me which I believe very often perplexes and disturbs men of serious and contemplative natures. David himself fell into it in that reflexion, 'When I consider the heavens the work of thy fingers, the moon and the stars which thou hast ordained; what is man that thou regardest him!' In the same manner when I considered that infinite host of stars, or, to speak more philosophically, of suns which were then shining upon me, with those innumerable sets of planets or worlds which were moving round their respective suns; when I still enlarged the idea, and supposed another heaven of suns and worlds rising still above this which we discovered, and these still enlightened by a superior firmament of luminaries, which are planted at so great a distance, that they may appear to the inhabitants of the former as the stars do to us; in short, while I pursued this thought, I could not but reflect on that little insignificant figure which I myself bore amidst the immensity of God's works ...

If, after this, we contemplate those wild fields of æther, that reach in height as far as from Saturn to the fixed stars, and run abroad almost to an infinitude, our imagination finds its capacity filled with so immense a prospect, and puts itself upon the stretch to comprehend it. But if we yet rise higher, and consider the fixed stars as so many vast oceans of flame, that are each of them attended with a different set of planets, and still discover new firmaments and new lights that are sunk farther in those unfathomable depths of aether, so as not to be seen by the strongest of our telescopes, we are lost in such a labyrinth of suns and worlds, and confounded with the immensity and magnificence of nature ...

There is no question but the universe has certain bounds set to it; but when we consider that it is the worth of infinite power, prompted by infinite goodness, with an infinite space to exert itself in, how can our imagination set any bounds to it? ...

Several moralists have considered the creation as the temple of God, which he has built with his own hands, and which is filled with his presence. Others have considered infinite space as the receptacle, or rather the habitation of the Almighty; but the noblest and most exalted way of considering this infinite space is that of Sir Isaac Newton, who calls it the sensorium of the Godhead. Brutes and men have their sensoriola, or little sensoriums, by which they apprehend the presence and perceive the actions of a few objects that lie contiguous to them. Their knowledge and observation turn within a very narrow circle. But as God Almighty cannot but perceive and know every thing in which he resides, infinite space gives room to infinite knowledge, and is, as it were, an organ to omniscience.*

---

* These excerpts were published in the *Spectator* as no. 420, *On the Pleasures of the Imagination*, 2 July 1712, and no. 565, *On the Nature of Man – of the Supreme Being*, 9 July 1714. Addison wrote often in this vein: cf. 519. *Meditation on Animal Life*, 25 October 1712, and 531, *On the Idea of the Supreme Being*, 8 November 1712. The latter derive from Locke and Aquinas.

Addison uses 'ether' for Earth's atmosphere, but 'æther' to describe the 'subtle fluid' once believed to permeate space and support the transmission of light waves.

Verse 8

8. Then did God move upon the face of the
Earth, and saw that it was lifeless and void.

IT is probable that Earth fell together about 4·6 billion
years ago, the consequence of particles being drawn
into a common gravity pool as the planet gradually constructed itself by accretion, sweeping up grains of stardust,
metal and silicates that had coalesced in the cooler outer
regions of the new sun's gaseous nebula.

There are opposing views as to how Earth came together. One has the iron core forming first, its powerful
gravitational field drawing in other particles, even micro-
planets, which then accreted around it. A second theory
has the planet coalescing as a homogeneous mixture of
metals and minerals, after which gravity pulled the heavy
elements down to separate them into Earth's metallic core.
Eventually this gravitational energy translated to heat,
melting the metals, making the core denser and thereby
increasing its gravitational attraction. It does appear that
Earth's metallic core began to separate out long before the
planet was complete. This differentiation into an iron core
and a surrounding rocky mantle has been fundamental to

the planet's geological architecture since the beginning. It is even possible that the core continues to pull iron from the mantle immediately above it, and that it is still growing and separating from the mantle of silicate-rich rock around it.

However the iron core formed, its separate existence since the planet's earliest years has made it a catalyst for change. Four and a half billion years later the core still sets in motion geological processes that fashion Earth's changing crust.

But if the core is the catalyst, the instrument of change is the thick mantle of silicate rock above it. Energy released by radioactive decay in this super-heated middle-Earth keeps rock near its melting point, so that it remains plastic and easily deformed. Convection currents set up in this cauldron of intense heat thrust columns of ductile rock toward the surface at the rate of just centimetres each year; slow, perhaps, but they rise inexorably, pushing their way up towards the crust.

Above this slow, vertical stream the solid crust forms an interlocking matrix of plates some one hundred kilometres (sixty miles) thick, supporting continents and ocean basins alike. The convection currents rise beneath these plates until they meet the resistance of the crust. Here they turn aside to flow outward beneath the crust, spreading away from their point of ascent. Eventually, as they travel horizontally beneath the crust, the currents begin to cool and their material to contract, until the rock-stream is borne

VERSE 8

down once more toward the fires of inner-Earth, thereby
completing the convective cycle.

Above all this, the actual continents are composed of
silicate- and alumina-rich rocks having a density just over
two and a half times that of water. They are light enough to
'float' on the denser surface plates underlying land and
ocean basins alike. Thus the continents ride, raft-like, on
crustal plates of heavier, magnesium-rich minerals. The
continents' cores of crystalline rock have been much broken and reduced by time, but, eons old, they survive.

The underlying plates do not endure. No material has
been found in sub-oceanic plates older than about 200
million years. They form, float and die, growing along one
fracture zone only to slough off along another.

According to tectonic theory, the mid-ocean ridges are
linear zones beneath which the ascending convection currents rise; and it is along these ridges that new rock for the
crustal plates is formed. The crust is here injected with
heated, plastic rock coming up from beneath, the plates
spreading apart as this new rock rides outward on the diverg-
ing currents from the upper mantle. Mid-ocean ridges are
the planet's growth lines along which juvenile rock forms, to
be pushed outward, away from the ridges, aging as it goes.
The spreading crust may thrust continents against each
other – India into Asia, their collision marked by the
Himalayas – or it may separate them, leaving ocean basins in
between. Where leading edges of crustal plates collide, one
will rise to form, for example, New Zealand; the other is

shoved under, sinking to oblivion as its substance is torn
away and reabsorbed into the rock-stuff of the upper mantle.

§ In 1996, three scientists announced in *Nature** that
they had discovered a tiny crystal from Earth's crust
embedded in a diamond. The tiny staurolite crystal would
have formed about ten kilometres down. But the diamond
in which it was trapped originated in the mantle, an esti-
mated 120 kilometres (75 miles) beneath Swaziland. The
staurolite must have been carried down in a descending
current or crustal plate – geologists call this subduction –
and preserved during the creation, under intense heat and
pressure, of the diamond. Then, over 200 million years
ago, a plume of superheated liquid rock (magma) shot the
diamond up into the crust. This microcrystal gave geologists their first proof that material formed in the crust can
descend to the mantle, and rise to the surface again.

§ Like shrapnel in a living body, some vast cold thing
is pressing on Earth's molten iron core almost 3,000
kilometres (1,850 miles) beneath India. In 1994, seismolo-
gist Michael Wysession noticed an object three times the
length of Australia while studying seismic traces reflected
from the boundary between the mantle and the core. He
noticed it because, being cooler than its surroundings, it
transmitted seismic waves rather faster. According to

* *Nature*, vol. 379, p. 153.

Wysession, this great slab of cooler rock is in just the right place to have once been the floor of the Tethys Sea.

The Tethys Sea formed about 230 million years ago, during the breakup of the proto-continent Pangaea ('all the land'). It is from this breakup that modern continents originated. When Pangaea pulled apart, the resulting cavity became the Tethys Sea. Thirty million years later the crustal plate supporting it slammed into Eurasia, and Tethys disappeared (all but the western end, which became the modern Mediterranean). The collision of the two great crustal plates forced the upper layers of Tethys' floor to rise, like wood shavings lifted by a carpenter's plane. These upper layers became incorporated into what is now the mountain ridge extending from the Alps to the Himalayas. The lower layers were thrust down into the mantle far beneath the continent. Could this ancient seabed have sunk 3,000 kilometres (1,850 miles) in two hundred million years, a speed of about one and a half centimetres (two thirds of an inch) a year? It may be. Eventually the core may heat it again, sending molten rock back to the surface, through which it may erupt to create a continent-sized tide of liquid basalt. Something similar happened in northern India 65 million years ago, leaving a vast 'flood basalt' formation, the Deccan flats.

§ As with life and the passage of seasons, so too with the fabric of our planet. Even rocks must die in order to renew.

Verse 9

9. Even the mantle of the Earth God found
wanting; for God breathed of the ethers that
were abroad throughout all the face of the
Earth, but among them was the breath of life
nowhere to be found, so the mantle was noxious upon it.

EARTH might have remained lifeless much longer if its
first atmosphere had contained free oxygen. In the
biblical sense, 'noxious' might well describe an atmosphere devoid of that life-sustaining gas, but during the first
stirrings of life on this planet the presence of oxygen would
have had precisely the opposite effect. It would have poi-
soned the first living things.

Through more than three billion years life evolved
under an atmosphere in which the proportion of oxygen
steadily increased. Green plants generated oxygen as a
byproduct of their photosynthesis. Life advanced as the
proportion of oxygen increased, and so evolved to cope
with it, later taking advantage of it. But the corrosive characteristic that makes oxygen essential to animal respiration
also gives it free rein to ravage living tissue.

The presence of oxygen would have prevented many organic molecules from forming in the early environment that eventually gave birth to life. That is not to say that life would have been impossible in an oxygen-rich atmosphere: it would have evolved differently.

The record of the past endures. Anaerobic bacteria thrive only in environments free of oxygen. The still more primitive methanogens depend for metabolism on their ability to oxidize hydrogen and reduce carbon dioxide into methane. These relics of the earliest life to conquer the planet now live in restricted environments, confined to deep muds or hot springs where they survive free from attack by oxygen. *Sic transit gloria.* But they have endured for more than three billion years, longer than any physical feature on the face of the Earth. Perhaps that is triumph enough.

Verses 10 and 11

10. And God passed over the face of the deep and smote his lightning staff upon the waters that even the ether was subdued and came together unto the lightning which he had wrought.

11. There God forged from the ether the very stuff of life in the womb of the seas; sugars, and amines, and nucleic acids created he them.

HOWEVER Earth was formed, it seems to be about 4·6 billion years old. To date, its oldest known surviving rocks, from Canada's Northwest Territories, were identified in 1989 as being 3·96 billion years of age.

The major miracle on Earth's surface is not that life came into being, but that it arose so early in the planet's history. In the first place, the planet had to settle into a condition hospitable to life; and then it was thought that the origin of life itself required an abundance of time. It was held as an article of faith that only time could unfold the complex miracle of evolution against all statistical odds to the contrary.

Several finds of ancient fossils refute that. South

Africa's Onverwacht Series of rocks yields alga-like cells dating back 3·4 billion years. And they are fairly advanced, obviously deriving from older, more primitive organisms. Australia offers similar fossils. These are not the oldest life-remnants discovered – Greenland samples date from 3·8 billion years – but heat and the pressure of ages has so altered older remains that their provenance is challenged. Figuratively speaking, the crust was barely cool before life stirred on its surface. For some people this shortened prelude demotes life from the miraculous to the mundane. For others, concision proves the miracle.

At the University of Chicago in 1953, Stanley Miller set out to reproduce Earth's juvenile atmosphere. Combining methane, ammonia, carbon dioxide and hot water, he bombarded this mixture with electrical discharges and ultraviolet light. Within a week the water turned a turbid red; amino acids, the building blocks of proteins, appeared as a slurry in his alchemist's brew.

Miller's classic experiment has long been bypassed by more recent thoughts on the subject. Ideas change, as capricious and fluid as the paths of evolution. But, if nothing else, Miller's work gave origin-of-life studies a significant boost. Decades later, the consensus is that the early atmosphere consisted largely of carbon dioxide and nitrogen. Two gases that Miller used in 1953, methane and ammonia, have been eliminated from the list of compounds likely in the early atmosphere. The sun – though it burned thirty per cent less brightly four billion years ago –

would have rapidly broken methane and ammonia into their constituent gases, nitrogen, hydrogen and carbon. In any case, it does not require large amounts of atmospheric hydrocarbons to generate organic molecules. Experiments similar to Miller's, using just carbon dioxide and water vapour with a trace of ammonia, also produce an amino acid soup under laboratory conditions.

Life-from-primordial-soup theories have fallen from favour lately. Water-based reactions tend to split molecules before they can grow and become more complex.

Atmospheric ammonia gives researchers problems, too. Significant amounts in the early atmosphere would have created a marked 'greenhouse effect'. That is to say, the sun's incoming short-wave ultraviolet rays would pass unimpeded through an ammonia-rich atmosphere, strike the surface beneath and be transformed into heat. The warmed surfaces would re-emit some of this heat-energy as long-wave infrared rays. A portion of the infrared would not escape back to space. Ammonia would absorb it, warming the atmosphere as a whole. Resulting high atmospheric temperatures on the young Earth might have been inimical to life.

For that matter, carbon dioxide also produces a marked greenhouse effect, but George Mullen and Carl Sagan calculated that a juvenile atmosphere rich in carbon dioxide would be warmed only slightly because strong radiation absorption bands were already saturated.

Perhaps the importance of the greenhouse effect in the

primitive atmosphere has been overstated. Earth's first thin envelope of gas probably amounted to less than one part in one hundred thousand (0·001 per cent) of our present atmosphere, and that represents the lowest theoretical density at which a significant greenhouse effect can set in.

One modern view is that Earth's early atmosphere, though thin and tenuous, may not have been greatly different from those of our celestial neighbours, Mars and Venus. Their atmospheres are rich in carbon dioxide released by early volcanic activity.

Perhaps a greenhouse effect was desirable, even essential, in Earth's early atmosphere. After all, four billion years ago the sun was thirty per cent cooler.

That fact makes origin-of-life scientists worry about something called the 'faint young sun' paradox. If our sun were to revert to its previous state, Earth's oceans would freeze. Chemist Jeffrey Bada and the durable Stanley Miller now propose that the early oceans may well have frozen, to a depth of at least three hundred metres. Chemical reactions associated with early life may have taken place in the water below, they say.

Other factors aside, some researchers still speculate that primordial ammonia was important to life's origin. Mullen and Sagan describe it as a 'useful precursor' to the first stirrings of organic chemistry.

Perhaps life's origin did need ammonia, but not necessarily from an atmospheric source. Instead, certain sand deposits may have played an important role in pre-life chemistry.

Two titanium-rich minerals, rutile and ilmenite, were originally locked into the fabric of igneous rocks, but time and erosion acted to lay them down as an important ingredient in many desert sands – and titanium dioxide is a potent catalyst in the conversion of nitrogen to ammonia. If this theory is credible, then the first stirring of life needed only shallow pools of water in titanium-rich sand for the necessary photo-chemical reactions to take place.

Bare rock has also been proposed as a suitable cradle for life. A team led by Leslie Orgel proposed in 1996 that the first self-replicating molecules formed on exposed minerals such as clay. Strings of nucleic acids grow better on bare clay surfaces than in water, the team found.

Pools, or bare rock. In either case life's chemistry evolved early, beneath a tenuous atmosphere more remarkable for its sparseness than for richness and diversity.

However life came to be, the very simplicity of the ingredients allows theorists to suggest that life had an extraterrestrial origin: the more spartan life's origin on *this* planet, the more likely that similar conditions occur elsewhere in the cosmos. Scientists have been trying to shift responsibility for the origin of life to some other planet since 1908, when Svante Arrhenius proposed that Earth's first settlers – a strain of interstellar bacteria – might have arrived from deep space. The theory was dusted off in 1997 after record-breaking television audiences watched NASA scientists hint that, just possibly, a piece of Martian rock found in Antarctica contained organic fossils. (The rock, it was suggested, had been ejected from the Martian surface by an asteroid impact.) If any organism could survive the temperature and radiation extremes of space, bacteria might. A species discovered in 1995 withstands a blast of radiation 3,000 times higher than a dose lethal to humans. The radiation shatters its chromosomes, but *Deinococcus radiodurans* fixes the damage from duplicate sets.

Scientists who visited Lake Baikal in 1990 have other suggestions for life's origin. The expedition discovered hitherto unknown species of clams, fish, snails and sponges clustered around mineralized hot water vents in a lake bed more than 400 metres deep. Their discovery rejuvenates an old theory that life, rather than requiring preconditions of sunlight and shallow water, originated in ocean depths, independent of sunlight, relying instead upon Earth's internal heat. If such were the case, it suggests that bacteria-like organisms made use of venting gases such as hydrogen sulphide as their source of metabolism, rather than depending on sunlight for photosynthetic reactions.

Whatever mechanism one prefers, by hinting at the humility of life's beginning, is science proving the miracle, or demonstrating the commonplace?

One other thought. Tests with experimental atmospheres tell us only that the groundwork is simple. Given the right conditions, amino acids in experimental apparatus will further combine to form spheres of highly complex proteins. More on that later.

In a sense, such experimental results no more resemble life than a pile of quarried stone resembles a gothic cathedral. The edifice has yet to be built. The most elemental living thing is a viroid, which is little more than a long ribbon of nucleic acid; but even a viroid's complexity is light-years ahead of the compounds in a scientist's jar. The next step is a pure virus, which wraps its nucleic acid ribbon in an envelope of protein. To attain that level of sophistication demands another great leap in life's grand design.

§ It is interesting to speculate about life's evolution from titanium-rich sand. We look back on the possibility of that event from the perspective of our species, which has learned to harness sand to extend our life-derived intelligence.

Sand, or silicon dioxide, is the stuff of which computer chips are made. Where life-given intelligence once evolved from a material slime, that intelligence can now look back and use its mineral origins in an attempt to extend its understanding of the world and whatever lies beyond. The cognitive force has made an appropriate full-circle turn of the wheel.

Verses 12 and 13

12. And God fashioned proteins, and complex large molecules, and to some he gave power of self generation.

13. And God swaddled each one in a cellular membrane, and he blessed them saying: Be fruitful and multiply, each according to his own kind.

To study life's origin is to collide with one indisputable fact: that, however it came about, life resulted from a falling together of appropriateness.

Whatever was appropriate for the origin of life *did* come to pass. Having said that, it's clear that too much has been made of the wondrous quality of life's origin. Some sense of wonder is essential. It represents the human capacity to gaze back into the long void of coincidence and happenstance so vital to the creation of animal intelligence and its predecessor, instinct. It was chemical instinct, after all, that provided the essential force through which certain organic chemicals fell together, reproduced themselves, and thus begat life. This is not to deny the possibility of divine guidance. William Cowper resolved the question to his own satisfaction with 'God moves in a mysterious way his wonders to perform.' On the other hand, Aleksandr Oparin spent a long and fruitful career studying the 'origin of life'. He introduced that phrase to the public, using it with complete objectivity and a scientist's dispassion. Oparin sought only to explain mechanism: questions touching on the possibility of divine motive and governance he left to others.

Indeed, to approach the origin of life armed only with a sense of wonder is to ignore the natural tendency of appropriate things to happen. Starting from first principles, simple organic chemical compounds have a tendency to organize themselves, and this is surely not much more than another step in the grand continuum of evolving systems. Chemical self-replication may be the source of organic intelligence and the beginning of life, but it stems from an older inorganic electrochemical intelligence that orders form and substance according to universal natural laws.

Cosmologists tell us that if our universe is destined to expand for ever, then its eventual fate will be entropy. That's to say, matter in all its diversity will eventually break down, to be replaced by the frigid and inert uniformity of a nothingness that is the ultimate simplicity. J. R. Newman described entropy as the universe's general trend toward disorder and death – a cosmologist's notion of apocalypse.

Until then it lives to harness its energies and its matter into an ordered hierarchy of relatively simple material and energetic forms. Thus it happens that the ninety-two elements in the periodic table combine into no more than six crystalline systems. This tendency to simplify things is Nature's own. For example, though laws of probability demonstrate that no two snowflakes can be identical, yet all of them conform to the strict hexagonal symmetry appropriate to the water molecules in their crystalline lattice. So it is with all Creation, living or inert. Its seemingly endless diversity represents sophisticated variation on the central theme of Nature's pragmatism, a pragmatism that ultimately renders everything into simple components.

§ Almost half a century of experimentation has shown that it is easy to produce complex organic chemicals in laboratory conditions. Great quantities of such compounds were formed on the juvenile Earth through the action of heat, radiant energy and electrical lightning discharge.

The usual residue left in experimental apparatus is a broth of amino acids, fatty acids and simple sugars. Sidney Fox, investigating pathways by which pre-biological chemistry gave birth to life, compared the many experiments involving laboratory-created atmospheres. He concluded that those yielding the largest number of life-compatible amino acids were those in which the experimental mixture of gases was subjected to steady, high heat in the presence of silica sand. Again we come back to sand.

At one time or another, similar experiments have given rise to each and every amino acid found in living things.

71

And when amino acids combine in such experiments to produce synthetic proteins (proteinoids), they do so in proportions similar to the amino acid ratios in living things. Furthermore, these amino acids are selective about how they combine with each other to form larger molecules. Their arrangement is in no sense random. Indeed, they show a high degree of chemical direction and arrangement, even at this humble molecular level. But then, so do snowflakes. Perhaps the sophisticated chemical arrangements that we find in organic compounds reflect nothing more than the evolution of inert intelligence. The recipe is eternal, for it combines its ingredients in the present and in the forever-after as it combined them in the past. The great falling together of appropriateness continues as it always has.

For a long time it was thought that life's chemistry was the end-product of such eons of random chance that no amount of investigation would reveal the paths by which it was first conceived. However, taking its lead from the self-directing nature of molecular chemistry, science has discovered that it is really quite simple to assemble primordial cell-like spheres of protein. George Wald pointed out that appropriate molecules could, often did, assemble themselves into microscopic structures with strong resemblances to living cells.

Early experiments along these lines involved collagen, a fibrous protein that binds animal tissues together. Briefly, when collagen-substance was precipitated out of chemical solution, new collagen fibres assembled themselves with the speed and efficiency of ice patterns freezing on glass.

Similarly, non-biological proteins can be induced to build themselves into complex cell-like beads called proteinoid microspheres. Here again, when protein molecules assemble themselves, the sequence of events is far from random. It appears carefully and chemically directed, with a few simple compounds undergoing specific reactions to create a structurally complex microsphere. For example, amino acids within the spheres sometimes show weak enzyme-like properties that assist and direct chemical reaction pathways. Sometimes spheres wrap themselves in a double wall analogous to the membranes of living cells: these barriers hold back large molecules but let small ones pass through. Finally, as if mimicking life, experimental microspheres have a 'tendency to participate in the reproduction of [their] own likeness,' writes Sidney Fox. They develop buds, some of which break away as new spheres, emulating the way in which yeasts and coccoid bacteria reproduce.

The pattern in which molecules assemble is consistent in every stage. It is non-random; it is chemically directed; and ultimately it is self-induced. At least, that is the careful illusion. Life, it seems, was the appropriate intention, outcome, and destiny of its own chemistry.

§ Setting aside proteins, consider the nucleic acids that give each living cell its genetic code. Nucleic acids determine the types of protein a cell will produce, thereby endowing it with the inherited characteristics it will pass along in its turn.

In 1953, James Watson and Francis Crick published a short article in *Nature* * suggesting a possible structure for deoxyribonucleic acid (DNA), the long-chain molecules which encode, store and decode the genetic information essential to living cells (excepting certain viruses, which are differently equipped).

Crick, Watson, and Maurice Wilkins shared a Nobel prize for discovering the elegant and intricate structure of the long-chain DNA molecule. This comprised two inter-wound spiral strands, each complementary to the other, made of interlocking phosphate and sugar compounds linked by nitrogen-rich units known as bases. Hydrogen atoms bond the two strands together. It seems appropriate that hydrogen, the first, the elemental form of matter, should bind the chemically-encoded storehouse of life's purpose and designs.

The double-strand model of DNA explains how genes replicate themselves so precisely. During cell division, hydrogen bonds between strands give way, leaving the two spiral chains unravelling from each other. But even as they come apart, new, complementary DNA-substance begins

* *Nature*, vol. 171, 25 April 1953, pp. 737-8.

forming on nodal points along each of the dissevered strands. The process soon results in two complete DNA molecules where there was just one before, each daughter-molecule composed half of new material and half of old.

The speed at which giant half-molecules of DNA can reconstruct themselves supports the contention that life's genesis 'appears to have an explanation of the utmost simplicity', as Fox puts it. From small to large organic molecules, self-assembly mechanisms were 'operationally simple at the same time that they were mechanistically complex'. In short, appropriate reactions happened selectively, with fluent speed.

The first proto-cell's hold on life must have been tenuous indeed. But for the extraordinary arrangement of atoms in its molecular lattice – and the unique volition of that lattice – that cell would have been as dead as the rock beneath it. Just a few atoms endowed one particular cell with the gift of 'life'. Imagine, one living cell on a dead and hostile planet. What an awesome prospect lay ahead.

It is interesting to compare the respective roles of proteins and nucleic acids in this process from a philosophical point of view.

The protein molecules in this first 'living' cell did more than allow it to reproduce itself. They embodied in this creature the definition of mortality. Until then, Creation had known only the constant certainty of natural laws masquerading as patterns of energy and mass. Then, all of a sudden, the self-chosen interactions of a protein molecule

75

gave rise to a mortal body – the soma of a cell – along with all that implies for Earth-bound mortality and the corruptibility of flesh.

A notion surfaces, borrowed from Christian burial rites. 'For this corruptible must put on incorruption, and this mortal must put on immortality.' The first organism's protein content represented its mortal body; the later development of nucleic acid long-chain molecules represented something else: they are life's tools for copying itself, passing down inheritance, and putting on immortality.

In one sense, then, a single-celled organism like a bacterium doesn't die. Splitting, it divides its life force into the bodies of two descendants, each endowed with the parent's genetic legacy.

Nucleic acid is the key to this inheritance, for in a real sense it carries the bright illusion of immortality along the ancient and continuing chain of life. Nucleic acid is the material tool with which life reaches out to touch eternity. 'As it was in the beginning, is now, and ever shall be, henceforth and forever more. Amen.'

Verse 14

14. And it came to pass that the ether and water with the light of the firmament gave some of those succour, and they grew abundantly in the seas that were about the Earth, and gave to her mantle their oxygen. And God saw photosynthesis that it was good.

THE first organisms to attain life must have sustained themselves by preying on other organic compounds. Aleksandr Oparin suggested that natural selection functioned even at that primal level.

In the absence of free oxygen, fermentation is the only process to allow that kind of predation. It breaks down organic compounds, rearranges their molecules and takes energy from the reaction. Yeast is the best known fermenting organism: it breaks sugars into alcohol and carbon dioxide, creating energy in the process. But fermentation is inefficient: the reaction releases but a small portion of the available energy, and its byproducts are toxic to the organism producing them. For example, alcohol kills the yeast that made it, so byproducts have to be got rid of if a reaction is to continue. This would have posed no great problem for single cells bathed in open water, but if fermentation were the only way in which living things could acquire energy, evolution would have been severely limited.

Somewhere in the moving stream of evolution, life's creatures found a better way of harnessing energy to suit their needs. The end result was a system – photosynthesis – whereby the energy of light is absorbed and stored as chemical energy, for later use by living cells. In its simplest form, green plants' tissues use light to convert carbon dioxide and water into a combination of oxygen, water and simple carbohydrates such as glucose or starch. Highly evolved green plants may store almost a third of the energy they absorb from sunlight in this manner. The logic of photosynthesis goes something like this: plant tissues capture the energy from sunlight to break the bonds between carbon, hydrogen, oxygen and other atoms. Breaking these chemical bonds releases the energy used by the plant. Some of it is put to immediate use rearranging atoms into more useful compounds. But a surplus remains, to be stored in chemical form.

The genesis of photosynthesis shows how preferred orientation was at work even in primordial chemistry. It represents the falling together of appropriateness. Bacteria, green plants, and much of the animal kingdom have evolved similar chemical solutions to the same problem. In fact, the chemical answer is similar wherever energy must be translated to meet life's demands.

The key to the success of photosynthesis is a pentagonal ring of carbon, hydrogen and nitrogen atoms known as a pyrrole. Four such linked rings constitute what is called a porphyrin structure, which constitutes the essential framework in many energy-conversion roles. The molecule responsible for photosynthesis in green plants is a porphyrin with a magnesium atom at its centre. A similar structure – evolved independently – provides energy synthesis in bacteria. If, instead of magnesium, a porphyrin's central atom is iron, then the product is haem (heme), the red pigment of animal blood. Other modifications to this basic structure give rise to molecules essential to respiration.

This single molecular lattice is central to much of life's chemistry, whether in bacteria, green plants or higher animals. The same structure has evolved at least twice, perhaps more often, to suit a variety of metabolic roles. And yet, each time, the path to that evolution looks to be as complex as the molecule itself.

A porphyrin molecule is the end-product of an interconnected chemical system needing its own synthesizing enzymes to make it work. One enzyme builds the components; another links them in a ring. The implication is that enabling enzymes had to evolve before the porphyrin, as if anticipating their tasks. Which came first? Which the cause, which the effect? Without the porphyrin the enzymes are useless, and without their specific functions their very existence makes no sense. The question shifts the 'chicken or the egg' debate back several billion years, removing it to a primary molecular level.

The paradox of the porphyrin's atomic framework may be the best example of the falling together of appropriateness. It is a perfect illustration of Nature's inherent logic. Such a logic must exist, for we often find it; we often see it; and we can describe its structure and its effects in some detail. But, as Thomas Aquinas put it, we can never discern its first, its uncaused, cause.

Photosynthesis had a dramatic impact on evolution. It allowed living things to manufacture chemical compounds from the chemistry around them. But there is more to its significance than using sunlight to build sugars. The same photosynthetic reactions can produce amino- and fatty-acids if nitrates, sulphates and other key elements are present.

At a stroke, photosynthesis endowed life with a large measure of independence over its environmental chemistry, at the same time adding free oxygen to the atmosphere. Even so, photosynthesis coupled with inefficient fermentation gives a cell only a tiny energy surplus. It was oxygen, the byproduct of photosynthesis, that eventually turned life's subsistence economy into one of surplus. To harness oxygen, life had to evolve another process – respiration.

The chemical logic of respiration goes like this: sugars manufactured by photosynthesis in a plant's leaves represent a fund of light-energy transformed into chemical-energy. This energy store becomes available to a plant when the compounds are broken down again by oxidation, through the process of respiration.

Respiration in plants, then, reverses the basic reaction of photosynthesis. Photosynthesis manufactures sugars (carbohydrates) in order to store them; then respiration burns them to liberate energy on demand. Using oxygen to burn carbohydrates releases a surplus of energy in the reaction, many times the energy released by fermentation. If photosynthesis gave living cells a measure of independence, it was respiration that liberated life's energies and allowed it to evolve and advance.

Verses 15 and 16

15. And God said: Let the waters of the seas be filled with life, and let there be green plants and animals in the depths therein, that all shall flourish abundantly and have increase each according to his own kind. And it was so.

16. Thus were the Archaeozoic and the Proterozoic eras even as the morning and the evening of the second day.

COLONIES of alga-like organisms thrived in sediments laid down 3·4 billion years ago, which later became rocks in South Africa's Onverwacht Series. Above them, in the Fig Tree Series, bacteria (*Eobacterium isolatum*) have been found dating to 3·1 billion years.

Single-celled organisms had already confronted and solved many basic environmental challenges by that early time, one quarter of the way through Earth's known history. Photosynthesis had liberated the first whiffs of oxygen into the atmosphere, setting the stage for efficient respiration. Henceforth, living things would no longer be strictly the creature of their atmospheric envelope, but rather the architects of it.

But although much of biology's infrastructure had slipped into place, life's fossil record shows no inclination to advance much during another two and a half billion years. Through much of that time, life's highest attainments were bacteria-like organisms, *monera*, and blue-green algae, which lack most of the organelles found in the cells of higher life forms. There is never a governing nucleus in these organisms, nor mitochondria to regulate metabolism. Perhaps the latter already enjoyed life as independent cells, but they had not yet been inducted into the structure of more complex creatures.

So, during a period representing almost two thirds of Earth's history, evolution appears to have been static while the cell itself evolved.

Verses 17 and 18

17. And the waters brought forth abundantly of moving creatures that have life through every part therein according to the word of God, for the waters gave unto them succour according to his commandment.

18. Life begat divers and myriad forms in the waters, even unto seaweeds that dwell on the rocks of the shore, and sponges that are within the deep.

ABOUT one billion years ago the pace of evolution began to quicken. Rocks reveal more complicated organisms; worm burrows, for example, and infrequently the fossils of their occupants. Creatures resembling jellyfish stranded on tidal flats, leaving impressions of their evanescent forms to harden into eternity.

But that was just the prelude. Give or take a few hundred thousand years, the fossil record explodes into a bewildering array of new creatures at the beginning of the Cambrian period, 543 million years ago. Life blossomed within a few million years, a mere tick of the clock compared to what had gone before. A definite date for the dawn of the Cambrian had long been subject to debate. It was established to the satisfaction of many paleontologists in 1993, when researchers from MIT and Harvard found volcanic pebbles embedded with early Cambrian fossils near the mouth of the River Lena in Siberia. The pebbles were rich in zirconium silicate, a precise source of radioactive dating. This yielded the figure 543.

For years scientists had been trying to come to terms with the fact that the first fossils representing most animal groups had appeared in a sudden blooming of life within a brief span of about thirty million years. The zircon-derived date contracted this window still more. It now seems that creatures from most of the great animal groupings – phyla – made first appearances in the fossil record early in the Cambrian period, not more than five to ten million years apart. Within this little epoch, Nature devised and came close to perfecting each and every body plan that has been successful to this day. The long, arduous winter of life's hibernation on earth blossomed within a few million years into an extraordinary, short, explosive spring.

Biology's Big Bang, as paleontologists describe this event, was not altogether unheralded. Fossils from the previous twenty million years reveal soft-bodied creatures: some may be colony-dwelling hydroids (related to jellyfish and corals); others, like *Dickinsonia*, may be segmented worms. Slowly but surely the fog is lifting from this pre-Cambrian epoch to reveal a marine environment rich in so-far-inexplicable life.

The veil of ages parted suddenly in 1998, when south China's Doushantuo stratum yielded embryos in the earliest stages of development. Some 570 million years old, these fossils of newly-fertilized eggs comprise just two, four, eight or sixteen cells. No one can say what creature these tiny balls of cells would have become, had they lived. But this find may reveal much about cell-level anatomy and embryo development in that 'long and lightless' time. We owe this discovery to the fact that calcium phosphate replaced organic tissue soon after it died. The same rocks yield fossil sponges too, making them the oldest known representatives of any multi-celled animal group still living.

As to the early Cambrian itself, the fossil record may play tricks: for the most part it preserves animals with hard, durable shells or chitinous exoskeletons. That does not explain why species bursting into the record of Cambrian life had a sudden need for armour. One theory suggests that newly-complex organisms had become more efficient predators, and that a shifting predator to prey ratio made armour a priority in the early Cambrian balance of terror.

The timing of this climactic life-surge depended on the evolution of porphyrin-derived respiration (see Verse 14), which in turn had to wait until the atmosphere accumulated enough oxygen to reach the 'Pasteur range', a table listing the oxygen pressure necessary to sustain life. The Pasteur figure varies from species to species. Furthermore, the oxygen pressure in the late Precambrian period is anybody's guess, but it has been suggested that the value was just one per cent of its present level. Although green algae were generating enormous quantities of oxygen, bacteria were using it almost as fast, breaking down tissues of algae in their turn.

§ The period leading up to Biology's Big Bang was so extraordinary that all fields of paleo-science regard the unexplained flowering of the Cambrian period as a supreme intellectual challenge. Among other questions: how did single-celled organisms give rise to those with many cells?

Higher plants evolved from amalgamations of single cells. (Multicelled algae appeared about 1·8 billion years ago.) As a group, algae encompass everything from simple motile cells to great trailing seaweeds. It is the intermediate forms, those made up of colonies of similar cells, that show how such transitions came about. In a plant such as *Volvox*, hundreds, even thousands, of identical cells embed themselves in the surface of a mucilaginous sphere, then behave like a single individual. When a *Volvox* swims toward light, all its cells cooperate like galley slaves driven by a collective will, their swimming motions carefully coordinated to reach their common goal.

Biologist Lynn Margulis has long championed her theory (now widely accepted) that the nucleated cell itself results from an ancient coming-together of bacteria for their mutual benefit. This accretion, this *symbiosis* on a bacterial level, did not happen all at once. It may have evolved over two billion years.

During that time, bacteria created and perfected life's essential biochemical systems: fermentation, photosynthesis, respiration, the fixing of atmospheric nitrogen into proteins. Bacteria devised these biochemical reactions, Margulis contends. All living things now depend on them.

Margulis' theory represents a falling together of appropriateness in biology not unlike the examples we found in organic chemistry.

And while bacteria were inventing life's chemistry they were also joining forces. In time the descendants of cohabiting bacteria would fuse into the efficient arrangement that biologists call the 'nucleated cell'. In Margulis' view, a series of fatal attractions among bacterial symbionts led them in time to become the efficient metabolic structure on which all higher life depends.

Mitochondria, the oxygen-burning power units in animal cells, and the chloroplasts that harness solar energy in plants – these are just two components of nucleated cells that Margulis describes as descending from independently-living bacteria.

If bacteria can combine to form cells, then cells can also combine into greater things. Many multi-celled creatures (*metazoa*) stemmed from amalgamations of single cells. Sponges certainly did. When certain living sponges are put through a fine sieve and completely broken up, the separated cells seek out their fellows, recombine, and organize themselves into their new creature, made of the same old parts. Higher animal life may well derive from a process of single cells amalgamating, but sponges are nowhere on the main tree of evolution. They are experimental creatures that lead to nothing beyond themselves. They are frozen in their evolution, and in time.

There is nothing unique about this ability of algae and sponges to order themselves into different forms and phases of existence. Many viruses can appear as inorganic crystals under the microscope; and slime-moulds can function either as separate animal-like cells that move around freely to forage, or they can come together as a single body to reproduce, like fungi.

We see how simple chemicals combine into organic compounds which then give birth to life; and how life itself has a tendency to organize and advance from simple forms to living things of great complexity. Even the most primitive organisms are often subject to a greater collective will. Some attribute such a tendency to biochemistry. Others call it God.

§ The fact that the 1920s were formative years for modern physics does not concern us directly. But intellectual fallout from quantum mechanics and relativity theory led to speculation about the nature of a Single Consciousness, a cosmic force of mind. Such a governing principle appeared to regulate each step of organization in the physical world. Sir Arthur Eddington ('the stuff of the world is mind-stuff') and Sir James Jeans attempted to explain this mystery in the realm of physics.

Meanwhile, A. N. Whitehead addressed the same phenomenon in biology. The study of biological systems on this planet confronts the handiwork of an apparent and omnipotent control mechanism on many levels:

· the falling together of appropriateness;
· molecules organize themselves at elementary levels;
· genetically coded material (principally DNA) replicates itself precisely through many millions of years;
· scrambled cells can organize themselves to re-create the original creature (sponges, slime-moulds, some green algae);
· undifferentiated embryonic cells migrate through an embryo and, taking appropriate positions, differentiate into organs – heart, lungs, nerves, brain – collaborating to become their final creature;
· creatures at many evolutionary levels (flatworms, salamanders) regrow amputated body parts.

Whitehead published his theory of 'organic mechanism' in 1926.* The time was ripe, he suggested, to introduce 'some new doctrine of organism'. By the 1920s, physicists had demolished the notion that matter was 'solid'. 'Science is taking on a new aspect which is neither purely physical nor purely biological. It is becoming the study of organisms,' Whitehead wrote. 'Organism' as he

---

* *Science and the Modern World*, Cambridge University Press, 1926.

used the term was not limited to living things. 'Biology is the study of the larger organisms: whereas physics is the study of the smaller organisms. This [Whitehead's treatise] is the theory of organic mechanism.'

Whitehead's treatise started from the perspective of sub-atomic physics. Applied to the organic world, he postulates a system of hierarchies, in which the electron constitutes the lowest level 'organism'. Above electrons, his hierarchy of organisms ranges up through atoms, molecules, cells, tissues, organs, and the creature that is the total body. Reactions within any given level of the hierarchy regulate themselves according to two sets of laws: those imposed within that level; and those imposed by the need to interact with the level above. Thus an electron is governed by laws binding atoms together. It is also subject to the laws by which atoms become predictable building blocks in molecules.

In Whitehead's scheme of things an entity (whether electron, cell or tissue) does what it must at its own level; and at the same time it obeys the demands imposed by the hierarchical level just above it. Whitehead's whole creature is built simultaneously from two different directions: from the bottom up; and from governing principle down.

Verse 19

19. Jellyfish also God caused to swim in the
seas that they should be brought whithersoever
the wind and the tide willeth, and to have habitation with the plankton both green plants and
animals that are within the deep.

SOMETIMES they are called coelenterates, a comment on
the fact that they consist of little more than a gut
cavity; sometimes cnidarians, for the stinging cells they all
possess. And their origins are obscure.

Between them, jellyfish and polyps comprise one of the
most enigmatic of animal groups. One thing is certain.
They are more advanced than sponges, which embody
most major cell types devised by living things. Jellyfish and
polyps added nerve cells, which sponges lack. With that,
Nature had virtually completed its repertoire of specialized
cellular building blocks. From here on, evolution deals
in changing form and complexity rather than basic cell
development.

Cnidarians represent an experiment with more than
nervous tissue. The trailing medusae of jellyfish and the
polyps of coral or sea anemones show little obvious affinity.

But they are coequals in an ongoing experiment making use of two distinct body types. Some species exhibit both, alternating from generation to generation.

The bell-shaped jellyfish *Aurelia* is a case in point. Its larvae settle on the bottom and become polyps, looking for all the world like coral animals. Then, in the next generation, tiny medusae break from the polyps and return to the free-swimming birthright of their grandparents.

Polyps are little more than inverted medusae, but most species eventually committed themselves to one configuration or the other and took on a final, irrevocable physical form. Jellyfish elected to become medusae, undulating through time and the oceans as ephemeral bags of ocean water held together by the membranes of a single life. They have left scant record of their presence or their passing, save for pressure marks where their soft bodies stranded on mud and died.

By contrast, polyps built structures to last. Collectively their skeletons have fashioned coral reefs into the most durable of all life's wonders. Nature's record is at once the most fleeting and the most enduring within this animal group.

Jellyfish, polyps and comb-jellies, those tiny pulsing spheres of clear jelly that besiege the summer coasts: where did they get their start? Are they the product of some aberrant sponge that suffered genetic failure? More likely they claim ascent from single-celled ciliates which developed a collective mind for grandeur, becoming architects of domes and shimmering tracery in the open seas. As a group they enjoy a perfect symmetry of form, a higher animal design originally cobbled together from single cells. Almost certainly they are the product of amalgamation, and, like sponges before them, they lead to nothing but themselves.

§ Some researchers challenge the neo-Darwinist view that natural selection is the sole force directing and guiding evolution. The shape of medusoid jellyfish larvae lends a sort of lazy logic to their argument.

Neo-Darwinists hold that Nature selects the most suitable (or kills the least fit) among random variations in a population, thereby shaping evolution's outcome. Neo-Darwinists see successful life forms as survivors in a never ending war between genetics and environment. For them, evolution represents triumph by that blind chance of appropriateness that we know as successful genes, the eons-old dictatorship of DNA.

A contending theory is that living things are subject to other constraints, namely the chemical and physical laws governing how they take shape. In this view, variations are not random; they are patterned on laws which, by virtue of universal and eternal application, are 'successful'. Furthermore, developing organisms – fertile eggs, for example – must constantly adjust groups of growing cells to accommodate other groups. Each juvenile organ yields or claims physical space and functions as it develops. The body of a growing creature is in a state of constant change as its body parts adjust to each other. It would be surprising, then, if living things, which can accommodate dynamic change among their own internal parts, were unable to respond to larger changes in the ever-shifting external world around them.

Which returns us to the shape of jellyfish larvae.

A drop of ink falling into clear water forms a flat disk. Then the edges of the disk droop. This results in a dome from the rim of which inky projections sink through the water at roughly equidistant intervals. At this stage the image resembles a juvenile jellyfish.

Physical laws determine the shapes of living things as inexorably as they determine those of inanimate ones. It may be that the jellyfish larva of old gained its shape because it offered the same passivity to support by water as the ink drop. Random variation goes hand in glove with evolution, but does it *drive* it, as neo-Darwinists suggest? The form of a juvenile jellyfish is as perfectly related to 'place' as the ink drop. All is related, at once enabling and enabled. The whole may determine the shape of its parts. We just can't tell how.

Verses 20 and 21

20. Then God caused worms to swim in the seas and annelid worms to crawl in the sands thereof.

21. And from some he made molluscs, and gave unto each one a shell and a foot to hold fast as the limpet the rocks of the shore, that these should become as the snails, even as the squid of the seas, and as clams sent he them forth.

IF single cells did not amalgamate themselves to create higher life forms, then they may have divided internally, or failed to separate after cell division.

Many single-celled animals propel themselves with rhythmic beats of thread-like cilia. They are known as ciliates, and some, like *Paramecium*, have many nuclei within their single cell. Such a creature would become multi-celled, a metazoan, if each nucleus partitioned off its zone of influence and thereby created many cells within the single body. It would then resemble the most primitive flatworms, to which some ciliates bear striking similarities. Flatworms have left few fossil traces. They are known in the record by their offspring, segmented annelid worms, and it is this transition which is fundamental to the ensuing flow of evolution.

When flatworms reproduce without benefit of sex, they simply split transversely into two halves, each of which regenerates the missing end. The old head produces a new tail, and the old tail a new head. But if the splitting process is interrupted, then the animal buds into segments, not unlike those of an annelid worm.

That is one possible origin for segmented worms like polychaetes, which swim in the open seas and burrow in sediments where their bodies and tunnels have been preserved. In time the annelids invaded dry land, and as earthworms began delving in soil instead of soft sand.

By the time annelids developed segments, part of the basic stock had already left the main line to form another major branch on life's tree. Molluscs do not begin to resemble worms except in their larval stages, for all that they are closely related. They are never segmented, except for a few primitive species like *Neopilina*, which hide their annelid segments beneath a mollusc's exterior.

Zoologist Ernst Haeckel once set out a biological principle to explain that any egg, in its quest to become an adult, passes through a series of larval or embryonic stages corresponding to the evolutionary pathway that gave rise to its species. For example, lobster larvae pass through successive stages in which they resemble, first, the young of coelenterates, then flatworms, annelids, and finally crustaceans. This progression suggested to

Haeckel that a larva ascends its own evolutionary tree as it grows.

Haeckel's principle made good sense for a time, and if it did nothing else it did stimulate research into hitherto ignored larval stages. Much of the invertebrate world gives birth to free-swimming ciliated larvae, many of which resemble each other. For a while all sorts of ancestral inter-relationships were postulated on the basis of these larval similarities.

Haeckel's Principle died a slow death when research showed that such larvae do not represent a single stage from which many groups of animals diverge. Their similar forms only infer that many species confronted common problems by evolving a common solution, arriving separately at similar answers. This process is known as convergent evolution. For example, some molluscs, like the octopus, have eyes similar to mammals'. Each evolved separately, without common antecedents, but in doing so developed identical forms to meet an identical need.

Invertebrate larvae, too, reached common solutions to common problems. Their shapes and functions are ideally suited to disperse their species through the length and breadth of the seas.

So, a larva's development does not reflect the evolutionary path of its species. But the need of larvae to meet demands made by natural selection did determine the evolutionary path taken by some species.

Three major animal groups fall under the heading of

molluscs: gastropods, comprising slugs and snails; cephalopods, like octopus, squid, and the extinct ammonites and belemnites; and bivalves, such as clams, oysters and pecten. Of these, gastropods most closely resemble the ancestral mollusc type. They offer convincing evidence that Haeckel's Principle, as well as the shape of living things, can be turned on its end.

Modern gastropods have turned some of their bodies' most fundamental organs around through a hundred and eighty degrees, so that they point the opposite way from those of ancient, fossil types. This torsion affects the whole body, including the nervous system. In modern species the two main longitudinal nerve lines cross each other. Oddly enough, such torsion seems to present modern gastropods with more problems than solutions, and it was not until zoologists considered the effects of natural selection with respect to the animals' larvae that this odd phenomenon began to make sense.

Growing larvae go through a surface-living phase, during which they manufacture protective shells. But these are effective only after the creatures go through their torsion phase – in order to tuck in their heads! A curious reversal of priorities: it would be more effective, surely, for animals to grow an adequate shell. But then they would sink. However, odds governing advantage and disadvantage in natural selection can work within tiny percentiles. Even the slight advantage conferred by an imperfect shell is magnified over many generations. Torsion came next, conferring a greater survival advantage. If, after all that, modern adults are somewhat discommoded – yet still they have life. The torsion of their adult bodies is the price their ancestors once paid for the survival of their larvae.

Molluscs remain eminently successful. More than seventy thousand modern species are known, and thirty thousand more from fossils. Molluscs evolved to fill every niche and role in the seas and, later, on land. Chitons rasped across wave-lashed rocks; snail-like creatures invaded the littoral, then the land; and bivalves burrowed and sifted their way through time and sediment as the ancestors of clams, oysters, and those free-swimming scallops with tiny molluscan eyes set like jewels around the edge of their mantles.

For some, the present is a shadow of the more glorious past. Great fleets of cephalopods once swam the seas, propelled backwards like squid, drawing their tentacles along behind them. All that remains is a profusion of straight-shelled belemnite fossils and the spirals of ammonites. Perhaps many of these long-departed ammonites migrated upward through ocean pastures to feed by night among surface plankton, but only the Pacific Nautilus remains as living evidence of this diurnal migration.

Squid survive, their ancient shells reduced to modern cuttle-bone; and octopus, with their uncanny 'human' eyes. Their eyes and ours show how Nature creates similar organs and forms many times over, across the living spectrum. Amidst continual divergence and change in her creatures, Nature returns to patterns that have been tried and tested many times before.

So it goes. Torsion and convergent evolution are but two smaller principles that serve to illustrate a greater one: evolution is Nature's expression of pragmatism.

Verse 22

22. To others he gave mantles of chitin, and jointed legs paired two by two, that of their offspring should come forth all manner of insects and spiders, and those which swim in the sea, and those which crawl beneath it.

T HE ancestral worm-stock was central to so many groups of animals that in retrospect it seems designed rather to digress than diverge.

The old worm-stock digressed again on the cusp of the Cambrian period, early enough to emerge into it as armoured creatures equipped with jointed limbs and definite mouth-parts, fully-fledged trilobites. The break must have come after ancestors acquired segments, for trilobites are segmented from stem to stern. Thin worm cuticle has been replaced by a flexible plate armour of tough chitin, and the limb-like lateral extensions of marine worms have been replaced by actual jointed limbs; hence the name arthropod – 'jointed feet'.

Trilobites prospered for two hundred and fifty million years before dying out. In the end they shared their fate with ninety per cent of marine organisms, wiped out in the massive extinction that closed the Permian period around 250 million years ago. (Several theories attempt to explain the worst extinction in the fossil record: an asteroid crash in the southern hemisphere; depleted oceanic oxygen levels; or a worldwide upwelling of carbon dioxide-rich water.)

Other arthropods persisted. The ocean filled with shrimps and crabs, lobsters and krill, crustaceans in such abundance and profusion that they became a central link in the food chain of the seas.

Later though, when crustaceans attempted to come ashore, they had limited success. Crustaceans have never escaped their need for constant moisture; but where they failed to gain a foothold on dry land, other arthropods in the form of insects would succeed.

One strange land dweller is as close as Nature ever comes to revealing its methods by leaving 'missing links'. *Peripatus* is a curious worm-like animal, whose smooth exterior belies internal annelid segments. The creature is an odd mixture of traits: clawed limbs like arthropods; a worm's excretory system; and a nervous system more primitive than a worm's. Yet *Peripatus* lives on, descended from the annelid line that gave rise to the arthropods.

The arthropod group must still have been young when the spider-like arachnids broke away in their turn. Arachnids must have diverged early on, because they were cheated out of several useful arthropod characteristics. Arachnids, from whom spiders descend, do not have the large compound eyes of other arthropods. They lack true jaws and sensory antennae, and their heads are fused to the thorax.

Some living arachnids go back practically unaltered to the Cambrian period. *Limulus*, the horseshoe crab of North America's eastern seaboard, belongs to one group with roots in the Silurian age, more than 400 million years ago. At that time, scorpion-like *eurypterids* were growing two metres long in shallow seas, and while they thrived on the seabed practically unchallenged, true scorpions came ashore. The oldest known land animal is a six-legged creature from this class. Found in Scotland, its fossil dates from 405 million years, laid down in rocks of late Silurian times.

True spiders came later, infesting the jungles and swamps that would eventually be preserved as the great coal beds of the Carboniferous periods, a time known in North America as the Mississippian and Pennsylvanian ages.

It used to be accepted that insects broke from the evolutionary tree much later. Their added sophistication seemed to require a longer gestation in the cradle of evolution than lower arthropods.

Not so. In 1988, the well preserved head of a bristletail insect was found in mudstone at Baie Gaspé, Quebec. Bristletails comprise the most primitive order of insects, similar to the modern household pest, the silverfish. But they are insects nonetheless, and the Quebec specimen with its large compound eyes is almost modern in form. The fossil dates from 390 million years, early Devonian time. For this creature to have made its debut in the early Devonian, insects may have evolved earlier. To date, the Quebec bristletail fossil is the oldest known trace of land animal life in what is now North America.

Verse 23

> **23.** And God saw life that it was good, and fashioned each as he desired, the starfish in the depths, and crustaceans that swim in the face of the deep.

IT would be a mistake to think of life's evolution as a progression of logical steps. Nature has launched valuable eccentricities into fascinating dead ends, some of which have outlived continents.

Starfish, sea urchins, and other spiny-skinned echinoderms achieved the most extraordinary adaptation when they branched off at a tangent from the evolutionary tree, becoming too specialized to lead to other things.

Their adults have the strict radial symmetry of polyps – or snowflakes for that matter – though for the most part echinoderms are pentagonal. It is the means to that end which is fascinating under a microscope, because echinoderm larvae start out bilateral, not unlike the larvae of that primitive stock leading to higher animals (see Verse 26).

When a larval echinoderm stops swimming and settles down, it starts to undergo a complete transformation. A perfectly sensible bilateral larva pushes and prods itself into a radial pattern like a five-pointed star. 'Front' and 'back' become notionally absurd, and in some species the adult animal has no preferred direction of travel whatever. The mutation that started this chain of events gave us the most improbable animals in the cabinet of creation.

Early echinoderms were as odd as modern ones. Fossils of crinoids, or sea-lilies, are found in many rocks laid down since the Cambrian age. The majority lived tethered lives, attached to the bottom by flexible jointed stalks that supported the creature by branching into the ocean floor. Some limestone rocks are composed of little else.

It is thought that these tethered crinoids gave rise to starfish later in the Palaeozoic era, for starfish are little more than crinoids that lost their stalks and fell, mouth downward, to the ocean floor. Once there, they began to crawl on tube-feet that had originally done double duty as respiratory organs and conveyers of food to the central mouth.

Were it not for their mundane larvae, echinoderms would seem foreign from our creation: they appear alien from most of life's tree. Their hydraulic tube-feet are an extraordinary accomplishment of natural engineering, unique to themselves; and the complex cantilevered mouth assembly of sea urchins, known to fishermen and scientists as Aristotle's Lantern, is a masterpiece of beauty and perfection in geometrical design. In a passage extolling the mathematics of form in the natural world, Sir James Jeans described 'The great architect of the universe … as a pure mathematician.' Pondering the mouthparts of a sea urchin, one can appreciate his point of view.

Verses 24 and 25

24. And it came to pass that seas covered much of the Earth, and the waters brought forth abundantly of coral within the deep. And there came up great pillars of fire upon the land, and hills of smoke from out of the waters that are about the Earth.

25. And God spake, saying: Every valley shall be exalted, and every mountain and hill made low. So God brake the land with wind and rain and caused the dust of the earth to return to the rivers and to enter once more within the deep. Then God caused hills to come forth from the waters and great mountains to come up also out of the seas, that the testimony of his might was graven on dry land, even as the shells that are fossil within the rocks thereof.

SOME eons ago we dealt with opposites. Or perhaps they were complements: complementary mass and energy states; opposing physical forces; and, in philosophical terms, opposed spiritual characteristics. Which brings us to mountain building, for Earth's highest mountains have often risen from its ocean troughs.

Great mountain ranges are made up of rocks that were laid down long before as shallow marine sediments. *Shallow* marine sediments. And yet the rocks making up the Appalachian mountains were once laid down to a depth of twelve thousand metres (forty thousand feet), *after* compression and consolidation. That amounts to a thickness of more than twelve kilometres (seven and a half miles). By comparison, the deepest trough in the present oceans, the Marianas Trench, is only eleven thousand metres (thirty-six thousand feet) deep. Hence the paradox: shallow water sediments in the oceans' greatest deep.

Henry and William Rogers made a definitive study of the Appalachian range in the mid-nineteenth century. They also answered the puzzle they themselves posed, offering a rational explanation for the enormous thickness of sediment involved. They reasoned that Earth's crust was sinking beneath thickening layers of sediment at about the same rate that sediment was being laid down. (In the Appalachian ranges the figure translates into one foot of sedimentation every seven and a half thousand years from the Cambrian period to the Permian, a span of nearly 300 million years.) The Rogers' explanation was partly correct. The crust *was* sinking, but the weight of sediment piled upon it was not sufficient to depress it. They had mistaken cause for effect.

Science has recognized another curiosity since Pierre Bouguer surveyed the Peruvian Andes in 1735. Bouguer discovered that mountains exert very little gravitational attraction with respect to their apparent mass. Similarly, Sir George Everest's survey of the Himalayas discovered that the whole Himalayan range deflected a plumb-bob by not quite five seconds of arc.

It took another century of work to reach a hypothesis that fits all known facts. In geology, good science may move little faster than the phenomena it studies.

Earth's innermost core, of iron, is solid despite its $7,300°K$ temperature by virtue of the enormous pressure confining it. Above the inner core, iron in the outer core is molten. Surrounding it, a 2,800 kilometre (1,800 mile) thick column of lighter minerals constitutes what is called the mantle layer. This, despite its density, is plastic. It flows. More precisely, it creeps (see Verse 8). Impelled by Earth's internal heat, this plastic rock forms immense convection currents that rise toward the surface at a rate of just centimetres a year, accelerating as they climb. Above the plastic mantle layer a thin shell of solid crust serves to support the continents and ocean floors – *and* protect the surface against the rising currents of plastic rock.

Whatever rises must fall. Having risen through the mantle layer, a convection current of plastic rock encounters the solid crust. Prevented from rising farther, it turns and moves parallel to the surface beneath the crust. As it goes, the plastic rock loses heat through volcanic activity or conduction. Cooling, it becomes denser. At last it falls back towards Earth's core.

Where such currents flow horizontally beneath the crust they exert a drag against it. And where two cooling convection currents meet and turn down, they pluck at the crust above them. This causes the crust to sag, producing a great trough at the surface, a trough that fills with sediment about as fast as it is formed.

Eventually all subterranean convection currents slow down and stop, and when they do they cease to pluck at the crust above. No longer tugged down, the crust relaxes: the trough then starts to rise, thrusting up a burden of sediment hundreds of metres thick, a burden long since compressed into rock. Imagine an emergent mountain range as a log held under water by an unseen force, then released to find its level. Just so, emergent mountains are thrust up until the system finds its point of equilibrium.

This explains why mountain peaks may be rich in fossil sea shells.

§ The essay after Verse 4 made mention of a gaseous nebula expelling two plume-like jets which then formed the spiral arms of a newborn galaxy. What follows here relates not at all, except that the action/reaction of jets exploding from opposite sides of a nebula brings to mind a factor connected with the great extinctions of animal life on Earth.

A growing body of evidence supports the theory that an asteroid impact near the Yucatan peninsula brought the dinosaurs and the Cretaceous period to a catastrophic end.

Less well known is the most drastic extinction-event of all time. It destroyed ninety per cent of marine species and perhaps brought down the curtain on the Permian period about 250 million years ago. Then, too, it is suggested that an enormous asteroid struck Earth, at a point now occupied by the southern Pacific Ocean between Tasmania and Antarctica.

That may be. Both events appear to coincide with equally catastrophic volcanic eruptions: the Yucatan impact coincides with a massive extrusion of basalt in India, the Deccan flats; the southern Pacific impact, if there was one, coincides with a similar volcanic rage in Siberia.

In both cases the volcanic eruptions broke from points diametrically opposite the supposed asteroid crash sites. A vast asteroid crashes into the surface, shatters the crust, the plastic rock of the mantle layer transmits the shock to the opposite point on the planet, splits the crust and emits catastrophic eruptions of lava. Action/reaction. Transmitted through the bowels of Earth. Interesting if true.

§ About this matter of convection currents in Earth's mantle:

Our planet's subterranean convection currents function, in a way, like the atmosphere above them. A worldwide system of atmospheric convection cells originates in the tropics, where the sun heats the equatorial zone, carrying warm air up and out to the higher latitudes in both hemispheres. Such is the great heat-engine driving the planet's climatic belts and weather systems.

Convection currents in Earth's bowels created, and continue to shape, the continents. It is thought that Pangaea, the original proto-continent, was formed when the lightest minerals, rich in silica and aluminium, were carried up by the rising currents to congeal, like froth on the surface of boiling soup. But what convection currents created, other convection currents later destroyed, breaking the proto-continent, driving Pangaea's pieces apart and acting as the once and future engine for continental drift.

Convection currents will continue to thrust up mountain ranges from ocean trenches. In the meantime the forces of frost and wind and water wear the rising surface down, trying in perpetuity to break the features of the land and return its substance to the seas. The result is that an air-born convection engine is for ever striving to pull down that which an Earth-born convection engine has striven to build up. So it goes. Things change in order to remain the same.

Verse 26

26. And God moulded of clay the first
creatures with backbones and set them apart,
that their issue should come forth abundantly
through all the estuaries of the seas, and begat
of their offspring the earliest fishes, like unto
lampreys that want even jaws.

EARLIER (Verse 23) we placed starfish and their allies in
an evolutionary backwater, but that is only partly
accurate. Their larval stages show that they diverged early
on from the branch of life's tree that gave rise to animals
with backbones.

Echinoderms and the first vertebrates were adept con-
tortionists. Echinoderms transformed bilateral larvae into
radial adults. The first creatures possessed of rudimentary
backbones also broke with precedent and reversed the
arrangement of major body parts. Invertebrates' main
blood vessels run along the dorsal side of the gut, and the
nervous system lies along the ventral side. Chordates and
vertebrates reverse that alignment.

At this point Nature poses another riddle, for it is possible that the whole vertebrate line once sprang from a

114

creature which, in a sense, never grew up. Adult sea-squirts appear to be among the least complex beings in the sea. Attached permanently to rocks, they are little more than open-ended sacs taking in oxygen and filtering food particles from the water. But their free-swimming ascidian larvae are fascinating because of the place they may hold in the evolutionary scheme of things. They have a chitinous strengthening-rod running the length of the back, and gill-like slits like those of other proto-vertebrates.

Zoology recognizes a condition known as neoteny, whereby certain creatures reach maturity and breed while their bodies remain in a juvenile stage. If such was once the case with the free-swimming 'tadpole' of an ancient sea-squirt, then in all probability the resulting creature gave rise to something like a lancelet, one of the most ancient of ancestral fish.

The lancelet is a small, translucent creature that swims and burrows in shallow seas, ignorant of its reputation in life's evolutionary tree. Its fame stems from the chitinous rod, known as a notochord, running the length of the animal's body to strengthen and support its internal struc-ture. The notochord anchors the animal's muscles better, letting them work more efficiently and thereby raising the lancelet's swimming speed. This simple addition effect-ively removed size limitations, encouraging further growth and diversity.

Chordates like lancelet discovered segmented structure as a way of adapting tightly muscled, undulating bodies to efficient movement in water. Through time and evolution, the notochord itself gave way to a segmented backbone of independent vertebrae, but the little lancelet is the creature that best illustrates the breakthrough. The animal is a miracle of innovative energy. Gills made their first appearance in chordates like lancelet, which uses them for straining food particles from sea-water. Eventually gills became the respiratory organs of fish, but one more evolutionary surge was too much for the lancelet. Its gills are still a device for straining food.

The first great divide in animal evolution is the ascent of multi-celled creatures from single-celled ones. The second is the development of backbones. Vertebrates are privileged, marked out for higher things. Many animal classes without backbones had already evolved their final forms and structures and were reduced to trying out new species by the time vertebrates started to progress from something like lancelets to something like lampreys, which were among the first jawless fish.

Verses 27 and 28

27. And God breathed of the mantle of Earth that it had oxygen among it, and he sent up scorpions from out of the deep to have habitation upon it, that they should be as a sign unto all that the dry land was good.

28. Thus it was that the Lower Palaeozoic era came to be as the morning and the evening of the third day.

IT is late in Silurian times, and the great explosion of invertebrate life lies buried more than 200 million years in the past. Some animal classes are well into a slow decline; others decayed or died out long ago.

We last looked at plant life a full three billion years back, at a time when algae were pumping oxygen into barren oceans and a thin sulphurous atmosphere. In the process, oxygen-producing algae were poisoning microbes unaccustomed to that searing gas.

Green algae prospered in intervening eons, evolving into everything from single motile cells to fronds of ocean seaweed. Meanwhile, spore-bearing plants must have come ashore long before they appear in the fossil record of the Silurian. Their first fossils are rootless, leafless shoots with spore capsules at the tips. Some of these swamp plants attained great size, with stalks eighteen metres (sixty feet) long and a girth of one metre (three feet) at the base.

Scorpions emerged into this landscape, making their first forays on land from shallow, swampy beaches. Large fossil colonies of related *eurypterids* have been discovered in association with some of the first swamp-dwelling land plants, preserved in rocks of what is now New York State. The animals may have come together year after year to breed in the shallows. That much is inference, based on the fact that the horseshoe crab, *Limulus*, continues the tradition, breeding once a year, precisely at the full moon nearest to mid-summer day. It is then that the highest tide falls near the longest day of the year. *Limulus'* innate sense of several circadian rhythms is such that, one month later, when the next full moon brings the next highest tide, young horseshoe crabs emerge from the sand and float back to sea in the receding water.

A brief digression. Evolution is more than the process by which animals and plants live and die and change their outward shapes and forms. It is seldom understood as such, but evolution is not in itself the process of change, but rather the process by which life as a whole seeks continuity through the vicissitudes of change. Continents have broken and drifted through thousands of kilometres and all sorts of climatic transformations; and for millions of years the biosphere has been awash in a constant state of flux.

But, through all that span of time and change, it is interesting to reflect that some of the most perpetual features in the biosphere are not composed of tissue, bone or even rock. They are abstracts, life's patterns of behaviour. Hence the parallel between Silurian *eurypterids* preserved in western New York's rocks and modern horseshoe crabs on New York's Long Island coasts. A pattern of behaviour may have been locked in place for something like 400 million years. Only the creatures themselves have been transformed.

We digress from the first fossil land plants in Silurian swamps and the protection those plants gave to the earliest terrestrial animals. Scorpions, and the millepedes that followed them ashore, remained within the shade and moisture of vegetation, a trait that millepedes have never seen fit to change. Other crustaceans, such as woodlice, migrated to land early on, and have always needed a source of moisture for their respirators to function. That factor has always limited crustaceans on land; their respirators still depend on a layer of water just molecules thick as a medium of gaseous exchange.

Atmospheric oxygen made this terrestrial migration possible in two ways. As the proportion of oxygen increased, it created a protective ozone shield in the high atmosphere which continues to defend living tissue against assault by ultraviolet radiation. And, of course, it became progressively easier for a greater range of animals to breathe. With an ozone shield aloft, and oxygen in the air, colonization of the land proceeded apace.

Verse 29

> **29.** And there came forth fishes in the estuaries of the seas, and some begat jaws and became as the sharks, others even as the bony fishes that are within the deep.

FISH are the most ancient of animals with true backbones. It used to be thought that they first appeared in the Ordovician period and went through a major evolutionary surge some 50 million years later. But, in 1995, a team from Birmingham University discovered shark scales and the remains of an early jawless fish in sandstone near Colorado Springs. Their find pushes back the earliest known fish fossils to 500 million years. 'What we have found,' says Ivan Sampson, 'is that fish went through a big stage of evolution soon after they first appeared.' It now appears that half a billion years ago the lineage of fish stemmed from something like a lancelet with evolutionary ambitions. Comparing living animals shows how changes came about.

In a modern lamprey, cartilage supports gills which have become efficient filters of oxygen rather than food. Their efficiency means that the number of gills is reduced to never more than six pairs. By comparison, lancelets may have over a dozen pairs of gills. Modern lampreys show how rudiments of the old notochord persist, but now there is a hint of a cartilaginous skeleton enclosing a well-developed brain in a definite head. A round, sucking mouth takes pride of place. There are no jaws. They come later. The earliest fish must have resembled lampreys in function, allowing for the fact that lampreys, to this day, lack fins. The evolution of fins in later fish is all important: they would evolve through time into lobes, into limbs, legs and wings, becoming eventually the arms of primates and humans.

First records of all this are lost. Fresh water fish left little trace in marine deposits, save for occasional shards of cartilaginous 'bone'. We have to wait for the Silurian period, when mud-grubbing jawless fish armoured themselves for defensive battle. Against what? *Eurypterids*, perhaps, those water scorpions whose relations played a significant role in the invasion of land. We know *eurypterids* and fish lived together in estuaries, and both declined together. *Eurypterids* were equipped to destroy all but the toughest armour; but armour is sometimes Nature's last line of defence for exhausted stock. Along with *eurypterids*, armoured jawless fish eventually perished too, leaving no legacy but their bones.

As long as fish lacked jaws they were condemned to grub through debris on the bottom, subject to attack. Thus the armour, making them correspondingly slow. Efficient jaws solved this problem, but Nature experimented with many strange mechanisms before fish evolved to their present forms.

Jaws gave fish the freedom to shed their armour and swim fast and free in open waters, but swimmers need a greater degree of hydrodynamic control than bottom dwellers. Fish responded by developing flaps of tissue which evolved into the paired pelvic and pectoral fins of all later species, giving fast-swimming bodies more balance and steering control. Those fins mean so much as evolution moves along.

An evolutionary impulse plucked fish from the mud and gave them paired fins with which to conquer the oceans. Environmental pressures would act upon fish again, turning fins into stout, fleshy lobes, releasing some fish from the oceans to set them free on the land.

Verses 30 and 31

30. And those there were which dwelt in rivers parched by the sun, and to these God gave lungs and stout fins of flesh that they might find pools to sustain them with life.

31. Thus crept they on dry land with primitive limbs, even as lungfish which dwell on the rivers' edge.

THE great age of fish lasted through Devonian time. Hard bone started to replace soft cartilage, except in sharks; and bony fish took the evolutionary lead. Together they diverged, they multiplied, they invaded the deep seas, and finally they conquered the land.

That circumstance probably owed a great deal to the vicissitudes of Devonian climate. There seem to have been marked seasonal fluctuations, with extended periods of heat and drought. The moon was closer, Earth's days were slightly shorter because the planet spun faster on its axis, and that combination made for rapid and extreme tidal fluctuations. In coastal waters this left fish susceptible to being stranded high and dry, vulnerable to drought, desiccation, and at the mercy of oxygen-depleted tidal pools.

Some early bony fish responded by developing a lung, an organ that persists to this day as a buoyancy control device, the swim bladder. They succeeded to the extent that twenty thousand species of fish make up forty per cent of all modern vertebrate species.

Long term evolutionary progress seldom rewards creatures that specialize rapidly and succeed within a particular niche. In evolution the long-term race goes not to the swift, but to more primitive, undifferentiated stock with staying power.

Thus it was that, while fish took over the oceans, two groups remained in the shallow half-world between land and sea. Lungfish survive to the present, gulping air in brackish pools, enduring months-long droughts as desiccated lumps of tissue preserved in dry mud. In the short term, some lobe-finned fish improved on that tactic. Their frail pelvic and pectoral fins evolved into tough lobes of muscular tissue. In time they would walk to water.

Lobe-finned fish succeeded, and they failed. They failed as shallow water fish, returned to the deep, and faded slowly from the fossil record, only to reappear alive as the deep sea coelacanth, *Latimeria*. As breeding stock, however, they were superb. Some creature from this line came to terms with both limbs and land, and as amphibians they came ashore to stay.

Verses 32 and 33

32. And there were fishes abundantly in great waters. And there came of their issue amphibian, which are as one upon dry land and in the rivers which are upon the land. And they multiplied in the sanctuary of the waters, and went forth upon the Earth.

33. And God made of their substance the frogs, even salamanders made he to be heirs unto them.

WHY should fish leave a watery world in order to fall on the mercy of hostile shores and uncertain weather? For the principle of natural selection to be effective, it must benefit each successive generation by conveying some advantageous characteristic. In hindsight it is clear that the ultimate effect of evolution on something like a lobe-finned fish was to put it ashore as an amphibian. We can describe such effects, but we should never describe them as goals. Evolution has no goals except stasis and survival. Apart from these, the record of evolution is one of completed effects, seen clearly only in retrospect.

The question to ask here is whether such an effect as

travel by land made sense from a contemporary lobe-finned fish's point of view. The answer is worthy of Alice's Wonderland: fish developed limbs to help them find water in order to escape from the land. In the process, some of them came ashore to stay.

The animal that made this successful transition to the land was, in all likelihood, neither a shallow-water lungfish nor a coelacanth, but a creature combining features of both.

Eusthenopteron, a Devonian air-breathing lobe-fin, must be close to the main line of ascent. A skull found in Quebec shows its nostrils connecting to the mouth cavity, characteristic of terrestrial vertebrates. Other specimens show similarities between *Eusthenopteron*'s large fin-lobes and the leg bones of later amphibians – indeed, of all later vertebrates. Fossils of this air-breathing fish bear a striking resemblance to those of an early amphibian, *Ichthyostega*, found in fresh-water deposits from Greenland.

Fossils discovered in Queensland led Australian researchers to claim in 1996 that the island continent could have been where fish first crawled out of the sea and where land-based vertebrate life began. Queensland University's Tim Hamley was quoted saying: 'Previously, the southern hemisphere 330 million years ago was thought to be too cold for amphibians to survive and most [similar fossils of this age] were found in Scotland.'

Wherever they first came ashore, amphibians landed, heavily dependent on water in which to lay eggs and keep themselves moist. More than 300 million years later they still appear to be creatures in culture shock, unable to cope completely on land, and reluctant to return completely to water. (In this respect they represent the vertebrate analogue of crustaceans, also reluctant drylanders.) The weak limbs of salamanders and newts still force them to move on land with the sinuous movements of fish. Their limbs do not propel them; they support them while the body moves forward, pushing against each limb in turn, much as a fish pushes itself forward by exerting pressure against the water around it.

Amphibians split into two groups early on. Tail-less frogs and toads grow to adults from juvenile forms in which vertebrae begin as loops of cartilage around the notochord. Only later do these loops turn to bone. That was the ancestral, the ancient development route. Most extinct species developed in that way, too.

But young salamanders, newts, and the legless snake-like caecilians have bony vertebrae from the start, never going through a cartilaginous phase. They make a curious group, isolated from the main; but, as if to underline their difference, Nature compensates for the loss of that developmental stage. They enjoy a peculiar privilege to an extent unparalleled in other vertebrates: they are endowed with the right to regenerate. Even adult salamanders can regenerate missing limbs. All but the youngest frog cannot.

The ghost of a theory reappears here to haunt the path of evolution. Haeckel's Principle of Recapitulation (see

127

Verses 21–2) has been dismissed as a tool for charting the evolutionary progress of a species from its past to its present, and yet the steps it illustrates are sometimes abundantly clear. This rite of passage is especially obvious in frogs. A tadpole grows from a fish-like phase to an amphibian in a few short weeks, losing its gills, growing lungs and limbs, and then absorbing its tail.

On the other hand, newts and salamanders seem always to seek regression to a more primitive stage. Many salamanders seldom develop fully. They retain their larval gills through life, even reproducing while still in a larval phase. The Mexican *Axolotl* always breeds in this fashion, and the North American Tiger Salamander may reproduce for several generations without troubling to grow up. This is a direct response to life in dry country. Retaining its larval gills ensures that just a touch of moisture on those gills will allow an animal to respire in a dry environment. The benefit of keeping those youthful gills is too great for a Tiger Salamander to cast them off in exchange for mere maturity. Such species reserve the adult body type for use in hard times. Only during a long drought, when its surroundings dry up, will an 'adult' juvenile throw off its perpetual adolescence, lose its gills, develop a lung and attain physical maturity. Under those conditions the salamander's hold on youth is reversed and Nature's discipline sees to it that the animal grows up.

With salamanders and their kin one wonders whether their failure to undergo the evolutionary step from cartilage to bone somehow arrested their development. They seem unable to compensate for something lost. But which of many seeming effects is the cause, and which seeming cause an effect? They are the less adaptable of the two amphibian groups, and everything about them begs to revert to earlier, more primitive creatures. They survive, and yet they bear the air of defeat, chained to their need for water.

In fact, all amphibians are fish out of water in some respects. To others shall go the spoils of further evolution, and the real conquest of the land.

Verses 34 and 35

34. And God caused plants to come forth from the earth. So there came up from the ground both horse-tail plants and the seed-ferns unto great height. And they breathed of the mantle and purified it.

35. Then God gave insects to be among the green plants, even the cockroach and the dragonfly brought he them forth from out of the earth.

PLANTS had a firm hold on land by mid-Devonian times. In what is now Wyoming, a shallow stream washed the prickly stalks of an upright leafless shrub, *Psilophyton*, into estuarine deposits. Remains of its stalks are crowned with spore capsules, and its shallow roots appear to have been used for support, rather than as the efficient foraging organs developed by later plants. At Rhynie, in Scotland, similar plants once grew four metres (twelve feet) high, their stems interwoven into a thick mat on the ground, struggling even then for a place in the sun.

Forests had taken root in what is now New York State by the end of Devonian times. Large tree-ferns stood twelve metres (forty feet) high, and many of the tall, spore-bearing lycopod trees bore leaves resembling dragon-scales on their trunks as well as on the branches. Horsetail plants appear for the first time; they survive to the present, but as shadows of their former glory. For over 200 million years, through the age of amphibians and reptiles, they were among the most abundant of plants.

These plants were forerunners of the great swamp forests produced in the Coal Age, tropical jungles that accumulated intermittently during sixty million years. It was a time of shifting lands, as shallow seas filled with river debris. Swamp vegetation built up, was inundated, and built up again. Evidence from coal seams laid down during those millions of years shows short-term fluctuations superimposed on long periods of stability.

Researchers used to wonder at the peculiar conditions of an age that permitted tropical climates to exist on most continents simultaneously. Now we know that the continents were still locked into one giant landmass, Pangaea, literally 'all the land' (see Verses 24–5). This was a time of uniformity, when successful species of animals and plants could conquer the entire habitable landmass. Spore-bearing plants of the Coal Age did so, just in time.

At the end of the Carboniferous period, possibly the cause of its demise, the proto-continent, Pangaea, started to break up, its torn-away continents wandering, like the galactic 'island universes' suggested by Kant.

Spore-bearing plants evolved through three billion

years to reach their culmination in the Coal Age. Simple unprotected spores were still the means by which plants reproduced, from the lowest unicellular alga, to mosses, ferns and horsetails. As plant life evolved, spores took on sexual attributes, combining in pairs to impart diversity to their descendants. In other species, new plants continued to germinate from single spores. This was the legacy of three billion years during which single-celled algae reproduced by simple budding.

Insects prospered under the umbrella of plant cover. Primitive flightless creatures like springtails are found in early Devonian rocks (Verse 22). Wings evolved later from plates on the first three thoracic segments. Proto-dragonflies took to the air, and were preying on other insects by the middle of the Coal Age. But rigid wings limit their owners to particular habitats and weather conditions. A new generation of insects equipped itself with wings folding back along the body. Cockroaches were among the first of these, and their remains are among the most common in the fossil record, perhaps because their remains were protected beneath their natural habitat of moist leaves.

Through the extraordinary gift of metamorphosis, some insects achieve in effect two lives. Early species like roaches never discovered this secret; their young are born as miniature versions of their parents, eating the same food. This did not stunt their distribution or success, but insects coming later confronted stiffer competition. To beat it, they evolved the advantage of differentiating their larvae from their adults. Leaf-chewing caterpillars do not compete with nectar-sipping parents for the same food source.

A Hindu legend illustrates the different realities within this greater whole. Ugly dragonfly larvae realize that when one of their number crawls up out of the mud and breaks through the surface of the pond it is never seen again. So they ask each departing nymph to report back, but not one of them returns. Instead, each nymph emerges from the pond to become a dragonfly, experiencing for itself the delights of flight over sunlit earth. Each new insect is firmly resolved to tell its fellows about the miracle awaiting them, but when a young dragonfly looks back at the surface of the pond it sees no mud, no other nymphs, no shafts of light shimmering through the water column. It sees a new and different world, one that throws back the reflection of an adult dragonfly, itself.

To borrow from a still more ancient source, 'The Search for Everlasting Life', in the *Epic of Gilgamesh*: It is only when the dragonfly larva sheds her skin that she sees the sun in all his glory.

Apparently insects prefer that dual reality. From what we can tell, only five per cent of Coal Age insects went through metamorphosis. Fully eighty per cent of modern insects cross that barrier to a second life.

Verses 36 and 37

36. So God gave unto the dry land life, and caused all manner of creatures to be upon it, and made of the dust both the plant that yields seed, and the plant that hath spores.

37. Thus were the Devonian and Carboniferous times like unto the morning and the evening of the fourth day.

THE forests of the Upper Coal Age took thirty million years to mature, to die, and to mature again in repeating cycles. Species were born in these forests, and died in them. The competition for light must have been severe, for the tallest trees grew straight to the sun before setting branches, in the manner of a modern rain-forest.

Lycopods, or club-mosses, dominated the forest. Up to thirty metres (a hundred feet) high, they branched near the top, their limbs supporting cone-like sporangia at the tips. Trunks as well as branches were often clad in scale-like leaves that left characteristic patterns on the wood.

The first seed-plants grew here among the lycopods. In time, competition from these newcomers might have destroyed the dominant spore-plants. As it was, continents broke and began drifting through climatic zones; land rose and fell, and other works of Nature conspired to kill the forests. From here on, the spores by which plants reproduced slowly gave way to seeds.

A spore may be nothing more than a genetically-coded cell, thrown to the elements by its parent with little protective covering or stored nutrition. It must establish itself quickly, or die. By contrast, when a seed is fertilized it goes some way to becoming an embryo within its own little world, replete with stored nutrition and a husk to protect it from the elements. Nothing is certain, and both seeds and spores may fall upon stony ground; but, from the moment when its parent ejects a seed, its odds of survival are better.

Verses 38 to 41

38. And it came to pass that God set reptiles on the Earth that they should have dominion over it in their time. Even as the dinosaurs sent he them forth in many and divers forms both great and small upon the land, and those there were which ran upon two legs, and those upon four even as the beast of the field.

39. Some sent he to rivers and the swampy ground, and yet others unto the depths of the sea as the ichthyosaurs, that they should return to the place whence life cometh and be abroad among the deep.

40. And God made great turtles to be abroad in the face of the waters, and crocodiles to be about the water's edge: and they were as one with leviathan, which delves in the earth and buries her eggs in the sand of the shore.

41. And it came to pass that God brought forth subtile serpents from out of the earth, that henceforth they should crawl on their bellies and eat of the dust all the days of their lives.

> So it was that God caused the serpents to go
> forth as every other creeping thing that creep-
> eth upon the earth.

WHILE seeds were replacing spores, reptiles were rising from amphibians. This makes for a curious parallel rise. Reptile eggs enjoy advantages over amphibian eggs comparable to those of plants' seeds over spores.

Embryonic reptiles, like seed-plants, develop within the security of their own package, replete with a source of nourishment. Young amphibians have no such comforts. The water in which amphibian eggs are laid becomes their mother, acting as the source of oxygen while carrying off their wastes. Water supports amphibians' eggs physically, protecting them against sudden desiccation, but it also enslaves them. Before vertebrates could truly conquer the land they had to break their ties with it. In time they did, starting with the reptile's egg.

A reptile's egg is an amphibian's pond in microcosm. It allows gaseous exchange, provides a large yolk for nourishment, and a sac to receive waste products. The embryo is swaddled against bruises and drought in a liquid-filled bag. Protected thus against the world, infant reptiles can become miniature adults within their little spheres. So it was that they emerged, over 300 million years ago, rising eventually to dominate animal life on their planet.

Until 1988 the earliest known reptile had been a 300 million-year-old fossil from rocks in Nova Scotia. Then that date was pushed back another forty million years with the discovery of an earlier skeleton, twenty centimetres long, in mudstone that was once the bottom of an ancient Scottish lake. Freshwater fish, spiders, amphibians and plants make up the otherwise unremarkable assemblage; but the reptile's presence among them, near the end of the Devonian period, has made paleontologists rethink the time scale for the evolution of vertebrate life on land. (This date may conflict with an Australian claim that the first amphibians came ashore 330 million years ago. See Verses 32 & 33.)

Setting aside this pioneer fossil, reptiles make their acknowledged debut in the coal swamps at the height of amphibians' supremacy. From the evidence of their bones they do not look like conquerors. Their clumsy splayed limbs are little better than amphibians', and it is a moot point whether these creatures had become true reptiles, or were waiting to do so. But, in time, reptiles gathered their limbs beneath them, or turned them to use as flippers or wings, or lost them, as snakes. Specific form matters little. Collectively they mastered the Earth for well over 200 million years.

Almost from their beginning, reptiles diverged into a bewildering array of shapes and habitats. Some of them had no sooner won their freedom from water than they returned to it, to an extent unprecedented even in amphibian forbears. Plesiosaurs reverted to flippers, with streamlined bodies and elongated necks, the better to chase fish.

Ichthyosaurs gave themselves back entirely to fish-like form, even to the extent of evolving dorsal fins. They are the reptilian analogues of sharks, which preceded them, and the mammalian dolphins which still lay far in the future. The external shape of these creatures is a testament to convergent evolution. Throughout the history of life on Earth, Nature has fashioned limbs, or organs, or entire species into similar forms and functions, in order to confront the challenge of a common environment, in this case water.

Then came the dinosaurs themselves, as diverse in their shapes and sizes as it is possible to imagine, many species perhaps warm-blooded, and among them the first creatures to stalk the Earth on two legs.

Reptiles took on many incarnations during their dominant years. They were at once rapacious tyrants, docile herbivores, and great horned beasts that trundled around in herds with their young protected at the centre. That much we can tell from their fossilized tracks.

Reptiles even conquered the air. Some would argue that they did it twice. Pterosaurs evolved to fill a niche of opportunity, ranging in size from a wingspan of a metre or so to over ten (a few feet to over thirty). They flew around for almost a hundred million years, defying, like bumble-bees, every theory of biophysics and aerodynamics, challenging human concepts of Nature's laws. Eventually they died out, eclipsed perhaps by the greater success of their cousins, the birds, which succeeded in the second great reptilian invasion of the air.

Verse 42

42. And to some he gave feathers and set
them apart as the fowl of the air that they might
fly in the open firmament of the heavens; even
as the birds of the air sent he them forth.

THE class of Birds consists of animals so essentially similar
to Reptiles in all the most essential features of their
organization, that these animals may be said to be merely an
extremely modified and aberrant Reptilian type.

Thus Thomas Henry Huxley in 1864, and his opinion has
been voiced many times in the years since. Some recent
work suggests that birds are related not only to the reptile
class as a whole, but specifically to dinosaurs.

It has to do with wishbones. Birds have them, but the
collar-bones that give rise to birds' wishbones are quite
absent in dinosaurs. This created some doubt that
dinosaur stock could evolve into birds. But collar-bones,
clavicles, have been discovered in remains of several early
reptiles, precursors to others known as coelurosaurs,
which resemble an early bird, *Archaeopteryx*. On the other
hand …

In 1997, biologists Ann Burke and Alan Feduccia disputed the prevailing theory that birds descended from the early group of running dinosaurs called theropods. It's a matter of fingers. Dinosaur 'fingers' correspond to the first three digits of the human hand. There is no question that they represent the first three: some early species show vestiges corresponding to fingers four and five. But when the biologists analyzed developing bird embryos they noticed what appear to be the beginnings of five digits. The outermost two disappear in later stages, suggesting that the 'fingers' in the wings of modern birds correspond to digits two, three and four.

Which, in a way, returns us to T. H. Huxley's assessment of avian anatomy in 1864. Researchers have been wondering since his day which three fingers birds possess.

There is little dispute that birds evolved from reptiles. It's just that some who oppose the bird/dinosaur link contend that birds and reptiles may be almost equally ancient; that both groups are cousins, as it were, related through a very early common ancestor to both.

§ A note about early birds:
It is not the oldest, and it may have been a dead end, but the primordial *Archaeopteryx* is by far the best known of the early birds. It is represented by six skeletons, the largest of which, the Solnhofen specimen, is little larger than a crow. *Archaeopteryx* was identified as a bird more

than a century ago from specimens discovered in the 1860s and sent to the British Museum and the Humboldt Museum in Berlin. In fact, when those two specimens were discovered, their existence was seized upon as evidence of the truth of Wallace's and Darwin's theories of evolution by natural selection. But the similarity between this particular early bird and contemporary dinosaurs is shown most clearly in the tale of some long-dead paleontologist's mistake: the Solnhofen skeleton languished for years, incorrectly identified as a bipedal, carnivorous dinosaur, *Compsognathus*. If *Archaeopteryx* was a dinosaur-like bird, *Compsognathus* was a bird-like dinosaur; and the similarity between the two is important evidence for those who see in *Archaeopteryx* a 'missing link'.

Archaeopteryx lived 147 million years ago. In 1984, two fragmentary skeletons of a creature named by its finder *Protoavis*, 'first bird', were found in western Texas. *Protoavis* lived 75 million years before *Archaeopteryx*, making it contemporary with early dinosaurs. Perhaps this should cause no surprise: some birds had attained nearly modern sophistication in early Cretaceous times, 130 to 140 million years ago, leading researchers to look further back than *Archaeopteryx* for the true proto-bird.

Paleontologist Hou Lian-hai believes birds might have been part of the scene before the end of the Triassic period (*c.* 210 million years). He also considers *Archaeopteryx* an evolutionary dead-end. In 1995, Hou discovered three late Jurassic bird species in layers of volcanic ash and mudstone from Liaoning Province. His 1995 discovery, *Confuciusornis*, had a modern-looking beak – unlike *Archaeopteryx* – but both species lacked the ridge-like keel along the breastbone to which modern birds' flight muscles attach. Presumably both were clumsy fliers. In 1996, though, Hou discovered another species, unnamed at this writing, the size of a warbler with a prominent keel and 'strong flight ability'. A second unnamed species had tail vertebrae fused into a single bone, a device that lets modern birds use their tail as a rudder in flight. Hou dates his Liaoning birds at about 142 million years, just before the Jurassic came to a close.

> That which is stated respecting the development and charac-
> ters of the amnion and allantois of the chick is true not only of
> all Birds, but of all Reptilia.

Huxley again, commenting on the similarity between eggs of two supposedly separate classes. They have in common the liquid-filled amniotic cavity, their equivalent of the amphibians' pond environment; they have a yolk sac attached to the ventral side of the embryo; and the allan-tois' life-support system doubles in both as a bladder and as a lung.

Birds are warm-blooded, an essential requirement for small creatures exposed to cold winds aloft. In addition, their circulatory system separates fresh oxygen-rich blood from depleted blood, using a four-chambered heart to pump it to flight muscles. It used to be thought that these were completely avian innovations, but many researchers believe that dinosaurs were warm-blooded too.

Reptiles dominated the Mesozoic era, changing their forms and their habitats, but never sacrificing their domination. Birds stayed small and evolved quietly, judging from the record, experimenting with flight until well into the Cretaceous period, when they successfully challenged flying reptiles for supremacy in the air.

How does such a creature as a bird develop within the theory of natural selection? Feathers are not difficult to explain. Chemically they are similar to reptilian scales. Furthermore, the smaller a warm-blooded animal becomes the more it needs an additional insulating layer to conserve heat.

But wings cry out for explanation. Why should a small running reptile set evolutionary changes in motion that would eventually turn its arms into its descendants' wings? Not to fly, certainly. It is of no possible advantage to a creature to know that twenty thousand generations hence its descendants will take to the air. If natural selection is to work, there must be immediate benefit to a creature or its offspring, if only by a tiny percentage. The first proto-birds might have used feathered arms like a shawl, to conserve heat. Next, wings would give lift and speed in a running hunt to capture food; later still, as aerofoils to assist proto-birds to glide among trees, much as flying-squirrels do. Delicate claws on *Archaeopteryx'* wings suggest that it might have clawed its way up tree trunks to launch itself into a glide. Modern birds whose wings are adapted to glide long distances over water do the same – Japanese sheerwaters, for example. Eventually came flight, and with it partial liberation from the dangers of the world below.

As a class, birds are remarkably uniform, their bodily design circumscribed by laws of aerodynamics. Exceptions show that some birds once had the ambition to fill the vacuum left by reptiles at the end of the Cretaceous period, some 65 million years ago.

Enormous flightless birds appeared on every continent at about that time; ostrich-like animals of extraordinary strength and size. In some cases they persisted right into the post-glacial period, but as predatory mammals appear in the fossil records the great birds disappear. The thousand pound 'elephant-bird' of Africa and Madagascar, *Aepyornis*, survived, and so did the smaller moas of New Zealand, until the arrival of the ultimate predator, man.

European art in the thirteenth and fourteenth centuries depicted birds as symbols of freedom and escape, careless of the wretched enslavement below. As the great, flightless birds died out, those retaining the power of flight multiplied and diverged, in numbers and success, freed in their own way from earthly concerns.

Verse 43

43. And God made of the reptile the beasts
of the field, and gave unto each a coat of hair,
and they were as mammals which give suck to
the fruit of their womb each after his kind, that
they in their turn should inherit the Earth, even
to the uttermost parts thereof. And God created great whales from among them, that they
should be their sign upon the deep.

EVERY evolutionary trend of early mammals was directed
to increasing their speed and rate of activity. In
defence, in attack, and in conquest of new environments,
mammals had to out-manoeuvre the dominant reptiles in
order to survive them; or to catch them, for that matter.
Many ancestral mammals were flesh-eaters, living by their
wits to catch their prey, and they needed larger, better-
organized brains combined with the advantage of speed to
sustain their ambitions.

Early changes show obvious improvements. Amphibians' undercarriages are grossly inefficient for sustained
movement on land. Reptiles' are better, and were further
improved in bipedal dinosaurs, but mammals rearranged

the ancestral splayed limbs in such a way that muscles propel bodies, rather than wasting energy supporting them.

Fossil skulls show that mammalian brains grew to become cognitive in a limited sense, not merely instinctive and reactive. Practically all this improvement took place in the cerebral hemispheres that originally handled sensory inputs. René Descartes' axiom 'I think, therefore I am' surely evolved along with cerebral capacity from a reptilian equivalent: 'My senses are aware, therefore my being reacts.'

Physical improvement alone might have been possible in the confined space of a reptile's egg, but cerebral and neural development needed a more sophisticated embryo, one demanding longer incubation than any store of egg yolk could allow. Amphibian and reptile eggs may be cast upon the waters or buried in the land, but each is a finite little biosphere, and its occupant can develop only as far as the constraints of its package permit.

In mammals, the womb replaces the finite egg, and in this safe maternal world the embryo is no longer restricted by a fixed supply of yolk. Instead, the placenta supplies the foetus's continuing needs while it develops. After birth, young birds as well as mammals have so much to learn and so much to coordinate that many species provide a parental bond for an extended period. Birds' instinct directs them to build nests and feed their young; mammals suckle their offspring and sustain them through growth and development.

While the great reptiles still lived, the dominant mammals were marsupials. Marsupials bear their young prematurely, forcing infants to crawl into a pouch on their mother's abdomen where the young continue to develop in a virtual incubator. But premature or no, marsupials lived up to Nature's demand for increasing intelligence. On that basis they prospered, branching into a full spectrum of roles in lands, primarily Australia, where isolation protected marsupials from the main thrust of evolution. The more efficient placental mammals evolved later, after continental drift had ripped the Australian tectonic plate away from other lands, segregating its marsupial inhabitants in isolated safety. Placental mammals came to dominate other continents, but not Australia. Indeed, marsupials coexisted in Australia with even earlier mammalian creatures, monotremes such as the duck-billed platypus and the spiny ant-eater, which continue to lay eggs.

Many of those creatures are now threatened. In direct competition with placental mammals, they retreat; and when they can no longer retreat, they die. The evidence of natural selection in mammalian evolution shows just how dominant is the maternal placenta and the umbilical cord connecting it to the embryo. The placenta and umbilicus have evolved into the lifeline of the mammalian class.

Verses 44 to 48

44. Of the dust God made cycad trees like unto palms, and the cone bearing tree he wrought also, whose seed is naked unto itself in the earth.

45. Then he created the flowering plant, saying: Let the earth bring forth grass and herb yielding seed, and the fruit tree yielding fruit after his kind, whose seed is in itself upon the earth. Even the fig and the willow tree brought he them forth.

46. Thus were the Mesozoic and Tertiary times even as the morning and the evening of the fifth day.

47. On the sixth day God looked out upon the Earth and saw all life that it was good.

48. And God blessed them, saying: Be fruitful and multiply, and fill the waters of the seas, and let fowl multiply in the air, and let the beast of the field and the creeping thing multiply each according to his kind. And it was so.

'LIFE's but a walking shadow, a poor player that struts and frets his hour upon the stage, and then is heard no more.' And that is how many major plant and animal groups appear in the fossil record. They come, they carve out a niche or they conquer, and then they disappear.

The larger reality is that, though a species may appear suddenly in the record of the rocks, its antecedents took tens of millions of years to rehearse for their hour upon Earth's stage. That is most true of plants, which appear and disappear in much less spectacular style than, say, trilobites or great reptiles. Living ghosts from Coal Age swamps still haunt us: giant club-mosses and horsetail plants are gone, but they live on in dwarfed species as the genetic legacy of an ancient time.

The rise of seed-plants is a case in point. Fossils of advanced and complex woody tissue have been found in Middle Devonian rocks, 240 million years before true flowering plants brightened landscapes of Cretaceous times. The first known seeds are not much more than sophisticated spores from an age when cross-fertilization was still a matter for wind and water to effect, without the help of insects.

Nevertheless, seed-plants originated in Devonian times, evolving eventually into two main classes whose classification depends on two associated factors: the manner by which female reproductive cells connect to the plant; and, whether or not a plant flowers.

Modern conifers represent the earlier of those classes,

the one in which female genetic matter is held 'naked' in cones, without the security of an ovary, hence the term 'gymnosperm', or 'naked seed'. There are no flowers. Nature is not profligate with favours, and in the Palaeozoic world from which the gymnosperms arose, no living thing could appreciate such abstract beauty.

Flowers came later, after flying insects took to the skies. That was when seed-plants evolved their second great class, 'angiosperms', the flowering plants. They came upon life's stage while mammals were still young; and while mammals were learning to cherish their young in a womb, angiosperms were developing ovaries in which to protect female sex cells and developing seeds. Impervious to weather, only grains of male pollen can violate the ovaries. From that it comes as no surprise that flowers occupy a prominent place in the symbolism of human sexuality. 'A little seed I spilled just in the centre, as I spread the petals to admire their loveliness, searching the calyx to its inmost depths.' Thus the thirteenth-century *Roman de la Rose**** expresses allusion in words for which there are parallels in many cultures. Flowers are such suitable symbols for erotic allegory.

So much for allegory. The reality of flowers is more banal. They are plants' tokens of good faith, their contribution to a symbiotic relationship with insects that has endured at least 130 million years. Many flowering plants

---

* *The Romance of the Rose*, © Florence L. Robbins & E. P. Dutton Co., New York, 1962.

succour insects with their pollen and nectar, while insects fertilize plants in their turn.

Therein lies the beauty of Creation; an interdependence of all life's forms woven together in an ever changing tapestry. We are the only species to enjoy a greater role within Creation than Shakespeare's Poor Player, for we alone have been given the part of surrogate director, to manipulate life's passion play on Earth.

§ Evolutionary biologists have puzzled over the processes by which new species originate ever since Charles Darwin suggested the mechanism of natural selection. Until recently, researchers preferred to believe that a new species evolved gradually from the old, that the transformation took place over extended time, countless generations, and resulted from many genetic mutations, each having a tiny effect. That view is shifting. It has become clear that, in certain cases, species division has resulted from a small number of genetic changes producing large effects.

In 1995, Toby Bradshaw and colleagues at the University of Washington (Seattle) analysed DNA from two species of monkey flowers growing in the Pacific Northwest. *Mimulus lewisii* is pollinated by bumblebees, *Mimulus cardinalis* by hummingbirds. Both species are perfectly adapted to their visitors. *M. lewisii* has pink flowers, visible to bees, with yellow streaks which guide them to a concentrated bead of nectar. *M. cardinalis* is red, a colour attractive to birds but not insects, and its petals form a tube leading to a pool of diluted nectar. In both cases the flowers' sexual organs are adjusted to brush the appropriate creature. Bradshaw and his colleagues concluded that just three distinct genetic changes had separated the two species, moving *M. cardinalis* to pollination by birds instead of bees.*

Speciation, it seems, need not be slow, only thoughtful.

The tree of the field is man's life.

– Deuteronomy 20.19

§ If a tree is man's life, what then is the nature of a tree? Not in and of itself, but from the human perspective.

Through history and the changing imperatives of human societies, there have been many *modes* of viewing trees. The perspective of time and social evolution changes the view; it did so in the past, and no doubt will do so in the future.

It is easier to review the record in the New World than the Old. The human time-scale, in so far as it deals with superstition and exploitation, is compressed in the Americas into recent, and recorded, history.

Michael Pollan, contemplating the ages of his New England farm at Cornwall, Connecticut, numbers several evolving stages in the human perception of trees. †

---

* *Nature*, vol. 337, p. 762.
† Michael Pollan, 'Putting Down Roots', *New York Times Magazine*, 6 May 1990, p. 38.

First came North America's aboriginal people, regular hunters in the forest that would become Cornwall. They took nothing but game, and left nothing behind but their trails. For the Indian, a tree was endowed with a soul, with senses; more than that, it had feelings as vital as those of any animal creature. That view of Nature is common to Animist peoples in all of the ages of man. We heard this voice before, in the essay supporting Verse 1: 'God, or Manitou, might speak as clearly from a rock, a hill, a burning bush, or the freshly-killed carcass of a hunter's caribou.' Suffice to say that the 'Indian tree', whether in Connecticut or Costa Rica, has sentient life, and its gatherings, its forests, are as thickly peopled by the spirits and ghosts of Animated Nature as any realm of the animal kingdom.

For which reason it had to die. In Pollan's progression, the Indian tree and its forests were followed by a Christian one, specifically, in New England, the 'Puritan tree'. To these worthy folk, the Puritan tree in its forested vastness represented not the Promised Land *per se*, but the promise of it. For Puritans, as Pollan quotes them, the New World forest was 'a hideous wilderness', 'wild and uncouth', a 'dismal thicket' where a body was liable to injury, ambush, death, or, what was worse, a falling away from Christ. To fell one tree, to fell an acre or to clear a township in the dark woods was a noble, Christian act.

Though later colonists acted upon secular, commercial principles to clear the wilderness, the impact was the same. Throughout the eighteenth century and well into the nineteenth, the 'Colonial tree' is either a commodity, providing barns, masts and barrel staves, or it is a nuisance, to be ripped out and cleared. If, to Puritans, deforestation defined godliness, to later colonists it represented progress, nay, a commercial imperative. Pollan records that when the land that is Cornwall was auctioned for settlement in 1738, the stipulation was that each settler must clear six acres or forfeit the title.

The nineteenth century saw the first shock of reaction against the manifest horrors of the Industrial Revolution and its 'dark satanic mills'. Romantics on both sides of the Atlantic rediscovered Nature, and redefined the tree. The 'Romantic tree' has gone full circle, described by Emerson in English as it was surely thought of many ages before, in countless Aboriginal tongues: 'In the woods, we return to faith and reason.'

The Romantic tree is more than ever with us, given a new urgency as the totem of our time, the symbol of mankind's treatment of the ecosystem. This symbolism is fiercely contested by commercial interests for whom trees remain a free good, to be exploited. The Colonial tree, you will recall, was to a degree just a nuisance, impeding agriculture. That role persists, to be sure, in reckless clearance by cattle-barons and landless peasants alike. In other parts of the world the inexorable march of clear-cut forestry neither knows nor seeks justification. Its imperative, and the vested interest of its workers, is commercial exploitation pure and simple, and that is deemed motive enough.

Pollan reminds us that our present debate over the role of 'Tree' in the community of beings is not new; it was taking place in North America a century and more gone by. Only the voices and the stakes are higher. Should our generation celebrate, as Walt Whitman did in 'The Song of the Broad-Axe', the achievement of loggers and frontiersmen? Or will we elect to fall in behind Henry David Thoreau, writing an ode to a pine tree felled by a logger's axe?

> A plant which it has taken two centuries to perfect, rising by slow stages into the heavens, had this afternoon ceased to exist.... Why does not the village bell sound a knell?

One of these points of view will prevail to determine the future for more than mankind. In all likelihood the planet itself is awaiting the result.

Verses 49 to 56

49. And God said: Let us make man in our spirit after our likeness; and he brought forth man upon the Earth and gave unto him a part of the spirit of God in his likeness that man should come to have wisdom in all things. For the thoughts of wisdom are more than the sea, and her counsels profounder than the greatest deep.

50. So God created man and gave unto him a part of the spirit of God that in wisdom he should have dominion over the fish of the sea and over the fowl of the air, and over cattle, and over all the Earth and every living thing that creepeth upon the earth.

51. And God blessed them and said unto them: Be fruitful and multiply and replenish the Earth. And God said: Behold I have given you every herb bearing seed which is upon the face of the Earth, and every tree, in the which is the fruit of a tree yielding seed, to you it shall be for meat.

52. And to every beast of the earth, and to every fowl of the air, and to every thing that creepeth upon the earth, wherein there is life, I have given every green herb for meat. And it was so.

53. And God saw every thing that he had made, and behold it was very good.

54. And God said unto man: Ask now the beasts, and they shall teach thee; and the fowls of the air, and they shall tell thee.

55. Or speak to the Earth, and it shall instruct thee; and the fishes of the sea shall declare unto thee.

56. Who knoweth not in all these that the hand of the Lord hath wrought this? In whose hand is the soul of every living thing, and the breath of all mankind.

WE take for granted the slow miracle of nature whereby a vineyard is irrigated and the water eventually becomes wine. It is only when Christ turns water instantly to wine that we are so utterly astonished.

– Augustine of Hippo

So saying, Augustine recognized the larger miracle; that Nature's processes are miraculous at any speed. We live in a miraculous world, but the miracle is lost to us because we live within it, as an integral part of it, and so it appears mundane.

In fact, the nature of the things about us has become so mundane that science can discover a rational explanation for almost anything in physical, in chemical, and ultimately in mathematical terms. Scarcely a process in the living world escapes explanation by material, rational science. Its thrust in the modern age has explained our life, our origins, our diseases, even the stirrings of our planet and our universe, in terms that are coldly and piously rational, and numerically sane.

Thus the sense of poetry and mystery by which a new leaf unwraps itself and grows, or the intricate development by which an egg becomes an embryo and then its creature.... All this we can explain, and rationally, for life's secrets have been rendered into a biochemical and electrochemical belief in the processes of DNA.

At this stage in our development we have learned to manufacture the genetic codes of life synthetically, inserting them into the deoxyribonucleic acid rings of lower organisms such as bacteria, changing them, creating in effect man-made life forms specifically designed to produce chemicals and pharmaceuticals at our bidding.

And as the minutiae of our knowledge increases, so the study of life passes down through scientific disciplines that deal with ever smaller fractions of the whole. Biologists give way to biochemists, biochemists to physical chemists, then to physical mathematicians and so on, until, at last, at the level of sub-atomic particles, life's study passes into the realm of pure mathematics. Ultimately the quest for life's secrets becomes as much the property of unified field theorists as those other considerations of theirs, gravity and the electromagnetic energy of light beams. Which still doesn't explain the miracle of life: it proves only that we have mastered the physical manifestations, the linkages, by which it chooses its wonders to perform.

There is danger here. It is that science runs on alone, indulged to operate within the closed circle of its self-important self-concern. In many fields the quest has become its own servant, enslaved to numbers and facts, seizing on the instant chaff of knowledge and leaving the real stuff of wisdom far behind. An intuitive understanding of the world and of its living things cannot easily be measured, and too often it is cast aside.

But still the search goes on; for the ultimate sub-atomic particle, and for the ultimate theory to explain the existence or the absence of that same particle. So once again we are dealing with Creation's ultimates and opposites; with forces and energies and particles that cancel each other out but refuse to disappear into nothingness and the void. Small wonder that the quest for the greatest secrets of life and the universe centres on the least of all natural things.

Put like that, the message and its meaning are an unfamiliar philosophical abstraction, but we can change the words a little, turning them into parable instead of physics. 'The kingdom of heaven is like to a grain of mustard seed

... which indeed is the least of all seeds ... but when it is sown, it groweth up, and becometh greater than all herbs, and shooteth out great branches.'

We recognize only the first, the intended meaning of the parable, which represents the sense of mission and success. The second sense is literal, and so it goes unrecognized for what it is. A part of the kingdom of heaven is bound up in every grain of mustard seed *per se* – and in every other sort of seed – for the biblical expression reflects the life-given force for fertility, development and growth.

So much is bound up in a single seed, or in a subatomic particle, for that matter. Proof, surely, that the potential for Creation – for the Becoming – is greater by far than the mere germ of its physical and numerical manifestations. As with the mustard seed, so too with all of its coeval living partners in this world, up to and including ourselves. If that lends all of us some measure of divinity, so be it. And if not, so be that too.

The final verses of The Becoming *are overleaf.*

Verses 57 to 62

57. But man heeded not the word of God that in wisdom he should have dominion over the Earth, but rather subdued it that it should be according to his will.

58. So man set himself over the spirit of God, even above the spirit of Creation which had brought him forth upon the Earth.

59. But the heavens and the Earth were finished, and all the host of them, and the morning and the evening were as the sixth day.

60. And on the seventh day God ended his work which he had made.

61. And he gave his Creation into the hand of man, that man should be as a steward unto all that the Lord God had made.

62. And on the seventh day God rested, and waited, that he should know how it would come to pass. So it was that man came to have power over Creation, either to magnify or to destroy.